Django 5 for the Impatient

Learn the core concepts of Django to develop Python
web applications

Daniel Correa

Greg Lim

Django 5 for the Impatient

Group Product Manager: Kaustubh Manglurkar
Publishing Product Manager: Bhavya Rao
Senior Content Development Editor: Feza Shaikh
Technical Editor: Simran Ali
Copy Editor: Safis Editing
Project Coordinator: Aishwarya Mohan
Indexer: Subalakshmi Govindhan
Production Designer: Ponraj Dhandapani
Marketing Coordinators: Anamika Singh

First published: June 2022

Second edition: September 2024

Production reference: 1250724

Published by Packt Publishing Ltd.

Grosvenor House
11 St Paul's Square
Birmingham
B3 1RB, UK

ISBN 978-1-83546-155-6

www.packtpub.com

To my mother, my ultimate hero.

– Daniel Correa

To my awesome wife for taking such good care of our family and children so that I could embark on my writing journey. She and our family are the very reason why I write books like this. Thank you so much, dear.

– Greg Lim

Contributors

About the authors

Daniel Correa is a researcher, software developer, and author of programming books. Holding a PhD in computer science, he also serves as a professor at Universidad EAFIT in Colombia. His interests lie in software architectures, frameworks, web development, and clean code.

I want to thank Greg for inspiring me to write books. Greg is one of the best programming book authors I have known. Thanks to my wife, family, colleagues, and friends for all the support. Finally, thanks to Miguel Sosa for his assistance with code revision, and thanks to the entire Packt team for the meticulous work to publish this book.

Greg Lim is a technologist and the author of several books on programming. He has taught programming in tertiary institutions for many years and places a strong emphasis on learning by doing.

I want to thank Daniel, my talented co-author; without him, this book wouldn't have been possible. Thanks also to everyone on the Packt team who helped us so much.

About the reviewers

Daniel Mitsuo Siena Hirata is a Brazilian full-stack engineer with over half a decade of experience, always up for solving problems, diving into games, and hunting for great food. He is on a journey through the ever-changing tech world, driven by curiosity and a love for learning new things. With years of Django experience and a solid grip on Python, he always add a creative touch to projects. He loves a good challenge and is always on the lookout for innovative solutions, making him a passionate and dynamic professional.

Nilton Pimentel is a Brazilian Python developer. He has more than 4 years of experience in the field of web development. He has developed many projects for companies using Django.

He is passionate about technology, soccer, games, music, and movies, and is always in search of knowledge and trying to evolve every day. His mission is to solve problems and deliver maximum value in people's lives using Python.

I'd like to thank my mother (Ivanir Pimentel) and my dog (Bolinha) for all their support and love throughout my career, and all the people who have helped me in some way to get here. THANK YOU!

Table of Contents

1

Installing Python and Django, and Introducing the Movies Store Application 1

2

Understanding the Project Structure and Creating Our First App 15

3

Designing a Base Template 29

4

Creating a Movies App with Dummy Data 47

5

Working with Models   59

6

Collecting and Displaying Data from the Database   73

7

Understanding the Database   83

8

Implementing User Signup and Login   95

9

Letting Users Create, Read, Update, and Delete Movie Reviews   117

10

Implementing a Shopping Cart System 137

11

Implementing Order and Item Models 157

12

Implementing the Purchase and Orders Pages 167

Preface

Django is a high-level Python web framework that encourages rapid development and clean, pragmatic design. Django is used for building modern Python web applications and it's free and open source.

Learning Django can be a tricky and time-consuming activity. There are hundreds of tutorials, loads of documentation, and many explanations that are hard to digest. However, this book enables you to use and learn Django in just a few days.

In this book, you'll go on a fun, hands-on, and pragmatic journey to learn about Django full-stack development. You'll start building your first Django app within minutes. You'll be provided with short explanations and a practical approach that cover some of the most important Django features, such as Django's structure, URLs, views, templates, models, CSS inclusion, image storage, forms, session, authentication and authorization, and the Django admin panel. You'll also learn how to design Django **model-view-template** (**MVT**) architectures and how to implement them. Furthermore, you'll use Django to develop a Movies Store application and deploy it to the internet.

By the end of this book, you'll be able to build and deploy your own Django web applications.

Who this book is for

This book is for Python developers at any level of experience with Python programming who want to build full-stack Python web applications using Django. The book is for absolute Django beginners.

What this book covers

Chapter 1, Installing Python and Django, and Introducing the Movies Store Application, covers Python and Django installation, and introduces the Movies Store application, showcasing functionalities, class diagrams, and MVT architecture.

Chapter 2, Understanding the Project Structure and Creating our First App, explores Django's project structure and app creation, and demonstrates how to use Django's URLs, views, and templates for creating pages.

Chapter 3, Designing a Base Template, explores how Django base templates can be used to reduce duplicated code and improve the look and feel of the Movies Store application.

Chapter 4, Creating a Movies App with Dummy Data, builds a movies app displaying a list of movies using dummy data.

Chapter 5, *Working with Models*, discusses the fundamentals of Django models and how to work with databases.

Chapter 6, *Collecting and Displaying Data from the Database*, discusses how to collect and display data from the database.

Chapter 7, *Understanding the Database*, shows how to inspect the database information and how to switch between database engines.

Chapter 8, *Implementing User Signup and Login*, discusses the Django authentication system and enhances the Movies Store application with some features to allow users to sign up and log in.

Chapter 9, *Letting Users Create, Read, Update, and Delete Movie Reviews*, enhances the Movies Store application with standard **CRUD (Create, Read, Update, Delete)** operations on reviews for movies.

Chapter 10, *Implementing a Shopping Cart System*, covers the use of Django sessions, and how web sessions can be used to implement a shopping cart system.

Chapter 11, *Implementing Order and Item Models*, explores how invoices work, and creates an Order and Item model to manage the purchase information.

Chapter 12, *Implementing the Purchase and Orders Pages*, creates purchase and orders pages, and concludes with a recap of the Movies Store's architecture.

Chapter 13, *Deploying the Application to the Cloud*, shows how to deploy Django applications on the cloud.

To get the most out of this book

You will need Python 3.10+ installed, pip, and a good code editor such as Visual Studio Code. The last chapter requires the use of Git to deploy the application code to the cloud. All the software requirements are available for Windows, macOS, and Linux.

Software/hardware covered in the book	Operating system requirements
Python 3.10+	Windows, macOS, or Linux
Pip	Windows, macOS, or Linux
Visual Studio Code	Windows, macOS, or Linux
Git	Windows, macOS, or Linux

If you are using the digital version of this book, we advise you to type the code yourself or access the code from the book's GitHub repository (a link is available in the next section). Doing so will help you avoid any potential errors related to the copying and pasting of code.

Download the example code files

You can download the example code files for this book from GitHub at `https://github.com/PacktPublishing/Django-5-for-the-Impatient-Second-Edition`. If there's an update to the code, it will be updated in the GitHub repository.

We also have other code bundles from our rich catalog of books and videos available at `https://github.com/PacktPublishing/`. Check them out!

Code in Action

The Code in Action videos for this book can be viewed at `https://packt.link/L3S8S`.

Conventions used

There are a number of text conventions used throughout this book.

`Code in text`: Indicates code words in text, database table names, folder names, filenames, file extensions, pathnames, dummy URLs, user input, and Twitter handles. Here is an example: "The db.sqlite3 file is the default SQLite database file that Django uses for development purposes."

A block of code is set as follows:

```
from django.contrib import admin
from django.urls import path

urlpatterns = [
    path('admin/', admin.site.urls),
]
```

When we wish to draw your attention to a particular part of a code block, the relevant lines or items are set in bold:

```
from django.shortcuts import render

def index(request):
    return render(request, 'home/index.html')
```

Any command-line input or output is written as follows:

```
python3 --version
```

Bold: Indicates a new term, an important word, or words that you see onscreen. For instance, words in menus or dialog boxes appear in **bold**. Here is an example: "For Windows, you must select the **Add python.exe to PATH** option."

> **Tips or important notes**
> Appear like this.

Get in touch

Feedback from our readers is always welcome.

General feedback: If you have questions about any aspect of this book, email us at customercare@ packtpub.com and mention the book title in the subject of your message.

Errata: Although we have taken every care to ensure the accuracy of our content, mistakes do happen. If you have found a mistake in this book, we would be grateful if you would report this to us. Please visit www.packtpub.com/support/errata and fill in the form.

Piracy: If you come across any illegal copies of our works in any form on the internet, we would be grateful if you would provide us with the location address or website name. Please contact us at copyright@packt.com with a link to the material.

If you are interested in becoming an author: If there is a topic that you have expertise in and you are interested in either writing or contributing to a book, please visit authors.packtpub.com.

Share Your Thoughts

Once you've read *Django 5 for the Impatient*, we'd love to hear your thoughts! Scan the QR code below to go straight to the Amazon review page for this book and share your feedback.

https://packt.link/r/1835461557

Your review is important to us and the tech community and will help us make sure we're delivering excellent quality content.

Download a free PDF copy of this book

Thanks for purchasing this book!

Do you like to read on the go but are unable to carry your print books everywhere?

Is your eBook purchase not compatible with the device of your choice?

Don't worry, now with every Packt book you get a DRM-free PDF version of that book at no cost.

Read anywhere, any place, on any device. Search, copy, and paste code from your favorite technical books directly into your application.

The perks don't stop there, you can get exclusive access to discounts, newsletters, and great free content in your inbox daily

Follow these simple steps to get the benefits:

1. Scan the QR code or visit the link below

https://packt.link/free-ebook/9781835461556

2. Submit your proof of purchase
3. That's it! We'll send your free PDF and other benefits to your email directly

1

Installing Python and Django, and Introducing the Movies Store Application

Welcome to *Django 5 for the Impatient*! This book focuses on the key tasks and concepts to help you learn and build **Django** applications quickly. It is designed for those of you who don't need all the details about Django, except for those that you really need to know. By the end of this book, you will be confident in creating your own Django projects.

So, what's Django? Django is a free, open-source web framework for building modern **Python** web applications. Django helps you quickly build web apps by abstracting away many of the repetitive challenges involved in building a website, such as connecting to a database, handling security, enabling user authentication, creating URL routes, displaying content on a page through templates and forms, supporting multiple database backends, and setting up an admin interface.

This reduction in repetitive tasks allows developers to focus on building a web application's functionality, rather than reinventing the wheel for standard web application functions.

Django is one of the most popular frameworks available and is used by established companies such as *Instagram*, *Pinterest*, *Mozilla*, and *National Geographic*. It is also easy enough to be used by start-ups and to build personal projects.

There are other popular frameworks, such as Flask in Python and Express in JavaScript (for more information on Express, see *Beginning Node.js, Express & MongoDB Development* by Greg Lim: `https://www.amazon.com/dp/B07TWDNMHJ/`). However, these frameworks only provide the minimum required functionality for a simple web page, and developers have to do more foundational work, such as installing and configuring third-party packages on their own for basic website functionality.

In this chapter, we are going to get acquainted with the application we are going to build, using Django 5, and get ready to develop our project by installing and setting up everything we need. By the end of the chapter, you will have successfully created your development environment.

In this chapter, we will be covering the following topics:

- Introducing and installing Python

- Introducing and installing Django

- Creating and running a Django project

- Understanding the Movies Store application

- Introducing Django MVT architecture

Technical requirements

In this chapter, we will use **Python 3.10+**.

The code for this chapter is located at `https://github.com/PacktPublishing/Django-5-for-the-Impatient-Second-Edition/tree/main/Chapter01/moviesstore`.

The CiA video for this chapter can be found at `https://packt.link/ygUpr`

Introducing and installing Python

Python is a high-level programming language (`https://www.python.org/`), created in the late 1980s by Guido van Rossum. The name Python comes from the creator's affection for the British comedy group Monty Python and not the "snake," as is commonly believed.

Python has an open-source license, meaning that developers can modify, use, and redistribute its code for free without paying the original author.

Python is characterized as a friendly and easy-to-learn programming language. Python can be used to develop a wide range of applications, including web development, data analysis, artificial intelligence, scientific computing, and automation.

For now, let's check whether we have Python installed and, if so, what version we have.

If you are using a Mac, open your Terminal. If you are using Windows, open Command Prompt. For convenience, we will refer to both the Terminal and Command Prompt as *Terminal* throughout the book.

We will need to check whether we have at least Python 3.10 in order to use Django 5. To do so, go to your Terminal and run the following commands:

- For macOS, run this:

    ```
    python3 --version
    ```

- For Windows, run this:

    ```
    python --version
    ```

This shows the version of Python you have installed. Make sure that the version is at least *3.10*. If it isn't, get the latest version of Python by going to https://www.python.org/downloads/ and installing the version for your OS. For Windows, you must select the **Add python.exe to PATH** option (to ensure that the Python interpreter can be accessed from any directory in the command prompt or *Terminal*), as shown in *Figure 1.1*:

Figure 1.1 – Installing Python on Windows

After the installation, run the command again to check the version of Python installed.

The output should reflect the latest version of Python, such as Python 3.12.2 (at the time of writing), as shown in *Figure 1.2*:

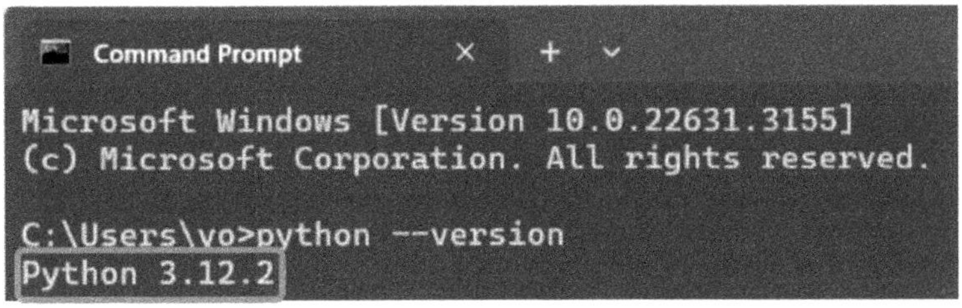

Figure 1.2 – Checking the Python version on Windows

Now that we have Python installed, let's move on to introducing and installing Django.

Introducing and installing Django

Django is a high-level Python web framework that encourages rapid development and clean, pragmatic design (https://www.djangoproject.com/). Django makes it easier to build better web apps more quickly and with less code.

There are several ways to install Django; we will use `pip` to install Django in this book. **pip** is the standard package manager for Python to install and manage packages not part of the standard Python library. `pip` is automatically installed if you downloaded Python from https://www.python.org/.

First, check whether you have `pip` installed by going to the Terminal and running the following commands:

- For macOS, run this:

    ```
    pip3
    ```

- For Windows, run this:

    ```
    pip
    ```

If you have `pip` installed, the output should display a list of `pip` commands, as shown in *Figure 1.3*:

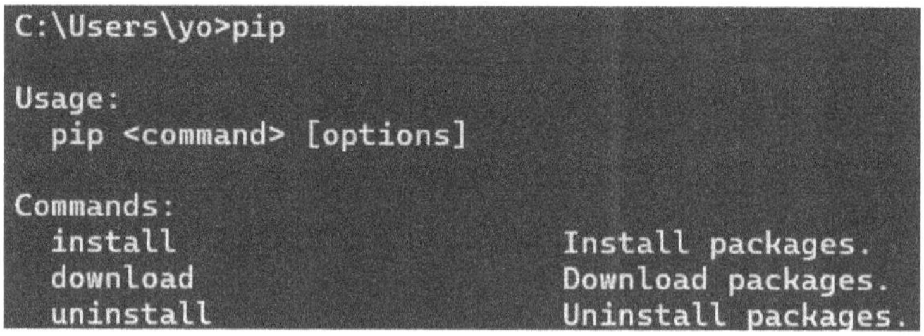

```
C:\Users\yo>pip

Usage:
  pip <command> [options]

Commands:
  install                      Install packages.
  download                     Download packages.
  uninstall                    Uninstall packages.
```

Figure 1.3 – Checking whether pip is installed on Windows

Next, to install Django, run the following commands:

- For macOS, run this:

    ```
    pip3 install django==5.0
    ```

- For Windows, run this:

    ```
    pip install django==5.0
    ```

The preceding command will retrieve the Django 5.0 code version and install it on your machine. Note that there may be newer versions available when you're reading this book. However, we recommend continuing to use Django 5.0 to ensure that the code in this book will function correctly. After installation, close and reopen your Terminal.

To check whether you have installed Django, run the following commands.

- For macOS, run this:

  ```
  python3 -m django
  ```

- For Windows, run this:

  ```
  python -m django
  ```

Now, the output will show you all the Django commands you can use, as shown in *Figure 1.4*:

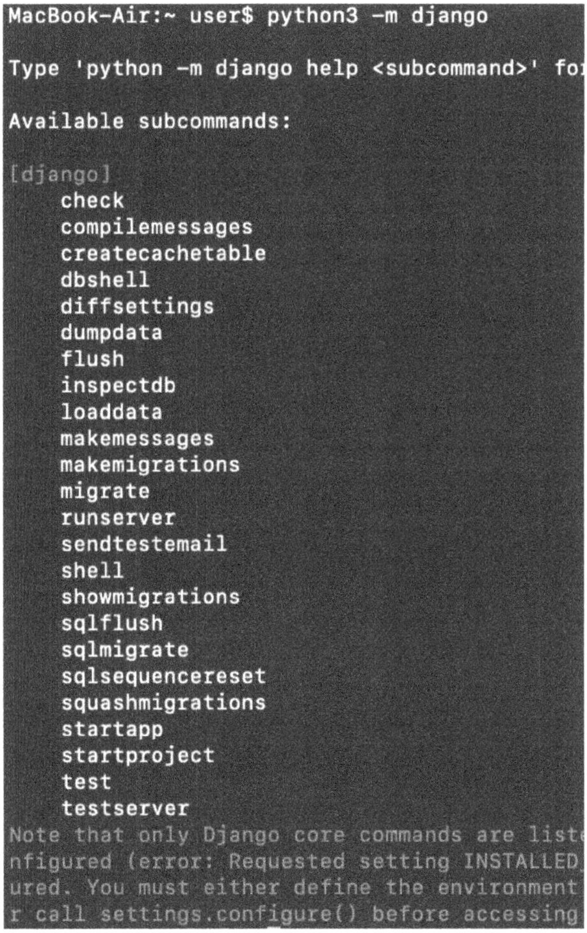

Figure 1.4 – The Django module commands on macOS

Over the course of the book, you will progressively be introduced to some of the preceding commands.

> **Note**
>
> It is also common to use **virtual environments** (such as the **venv** module) to manage your Python and Django projects and dependencies. For now, we will not use venv to get started quickly on Django. We will learn how to use and configure venv at the end of this book.

We have all the tools we need to create a Django project. Now, let's move on to doing that.

Creating and running a Django project

Now that we have Django installed, we are ready to create our Django project.

There are several ways to create Django projects. In this book, we will use django-admin. **django-admin** is Django's command-line utility for administrative tasks. It provides various commands to help you create, manage, and interact with Django projects, applications, and other related components.

In the Terminal, navigate to the folder where you want to create your project and run the following command:

```
django-admin startproject moviesstore
```

This will create a moviesstore folder in your current directory. This folder contains our Django application code. We will discuss its contents later. For now, let's run our first website on the Django local web server.

In the Terminal, run the cd command to move into the created folder:

```
cd moviesstore
```

Then, run the following command:

- For macOS, run this:

  ```
  python3 manage.py runserver
  ```

- For Windows, run this:

  ```
  python manage.py runserver
  ```

When you run the aforementioned commands, you start the local web server on your machine (for local development purposes). There will be a URL link – http://127.0.0.1:8000/ (equivalent to http://localhost:8000). Open this link in a browser, and you will see the default landing page, as shown in *Figure 1.5*:

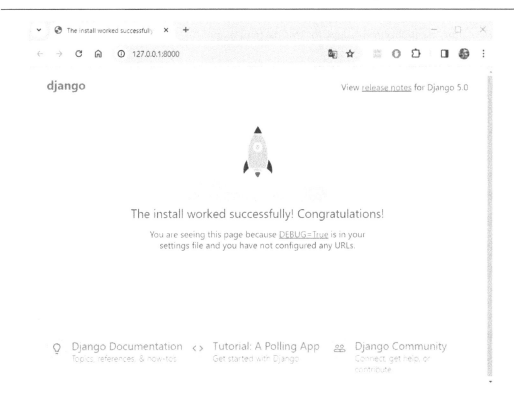

Figure 1.5 – The landing page of the Django project

This means that your local web server is running and serving the landing page. Sometimes, you will need to stop the server in order to run other Python commands. To stop the local server, press *Ctrl + C* in the Terminal.

We executed our first Django project successfully. Now, it is time to introduce the application we will develop in this book.

Understanding the Movies Store application

The use of running examples is a prevalent approach found in programming literature. The running example serves as a means to illustrate the principles of a methodology, process, tool, or technique. In this book, we will define a *Movies Store* running example. We will revisit this running example throughout the book to explain many of the Django concepts and elements in a practical way.

The *Movies Store* will be a web-based platform where users access information about movies and can place orders to purchase them.

Now, let's outline the application's scope for this particular app:

- The **Home page** will feature a welcoming message.

- The **About page** will provide details about the *Movies Store*.

- The **Movies page** will exhibit information on available movies and include a filter to search movies by name. Additionally, users can click on a specific movie to view its details and post reviews.

- The **Cart page** will showcase the movies added to the cart, along with the total price to be paid. Users can also remove movies from the cart and proceed with purchases.

- The **Register page** will present a form enabling users to sign up for accounts.

- The **Login page** will present a form allowing users to log in to the application.

- The **Orders page** will display the orders placed by the logged-in user.

- The **Admin panel** will encompass sections to manage the store's information, including creating, updating, deleting, and listing information.

The *Movies Store* will be developed using Django (Python), with a SQLite database and Bootstrap (a CSS framework). Further details about these components will be covered in the forthcoming chapters.

In *Figure 1.6*, you'll find a class diagram outlining the application's scope and design. The *user* class is depicted with its associated data (such as an id, username, email, and password) and is capable of placing *orders*. Each *order* consists of one or more *items*, which are linked to individual *movies*. Each *movie* will possess its respective data (including an id, name, price, description, and image). Lastly, *users* have the ability to create *reviews* for *movies*.

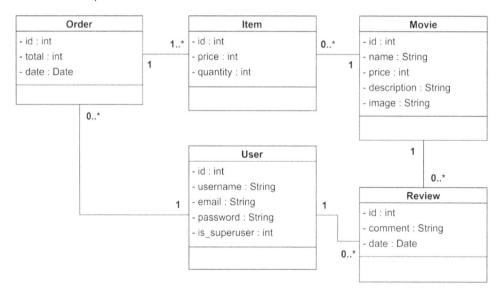

Figure 1.6 – The Movies Store class diagram

This book does not delve into the intricacies of class diagrams; hence, we won't elaborate on additional details within the diagram (you can refer to this link for additional information about class diagrams: `https://www.visual-paradigm.com/guide/uml-unified-modeling-language/uml-class-diagram-tutorial/`). As you progress through the book, you'll notice the correlation between code and this diagram. Serving as a blueprint, this diagram guides the construction of our application.

> **Note**
>
> Creating a class diagram before commencing coding aids in comprehending the application's scope and identifying crucial data points. Additionally, it facilitates understanding the interconnections among various elements of the application. This diagram can be shared with team members or colleagues for feedback, allowing for adjustments as necessary. Due to its nature as a diagram, modifications can be implemented quickly. Otherwise, once the project has been coded, the cost of relocating data from one class to another increases significantly. Check the following statement from the book *Building Microservices* by *Newman, S.* (2015): *"I tend to do much of my thinking in the place where the cost of change and the cost of mistakes is as low as it can be: the whiteboard."*

Based on the previous scope, we will build a *Movies Store* app that will allow users to view and search for movies, as shown in *Figure 1.7*:

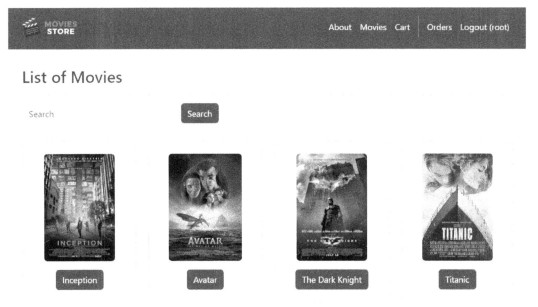

Figure 1.7 – The movies page with search functionality

Users will be able to sign up, as shown in *Figure 1.8*:

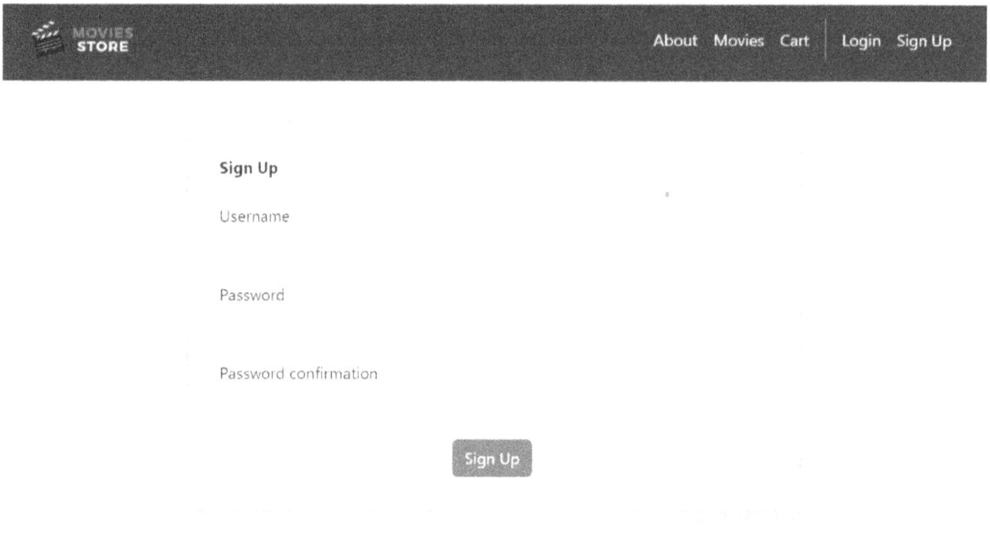

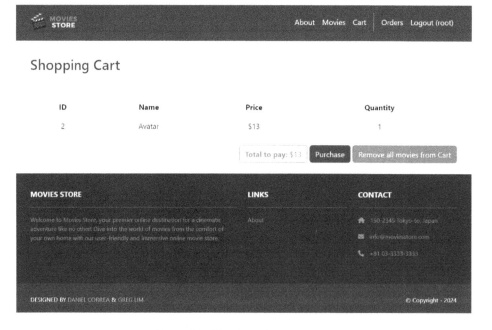

Figure 1.8 – The Sign Up page

Users will be able to log in, add movies to the cart, and make purchases, as shown in *Figure 1.9*:

Figure 1.9 – The shopping cart page

Users will also be able to list, create, edit, and delete movie reviews, as shown in *Figure 1.10*:

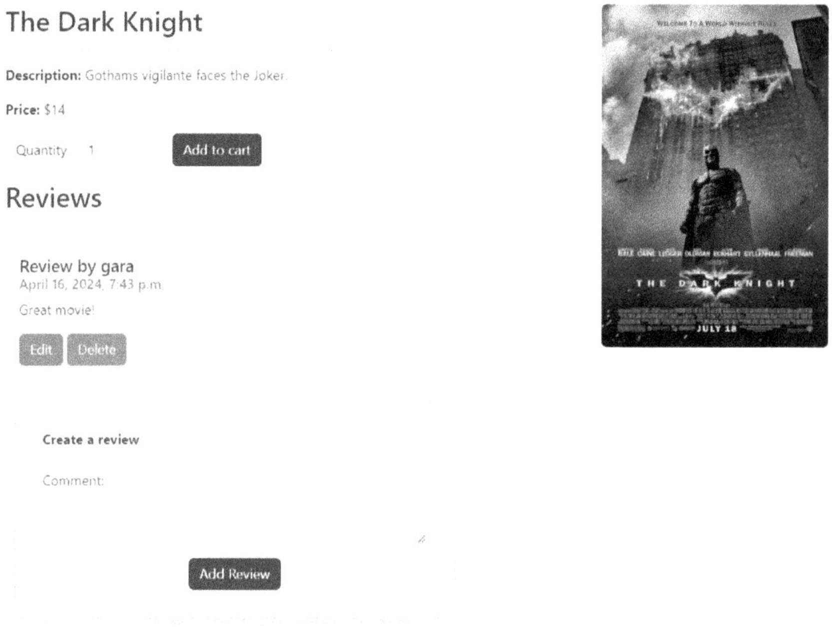

The Dark Knight

Description: Gothams vigilante faces the Joker.

Price: $14

Quantity 1 Add to cart

Reviews

Review by gara
April 16, 2024, 7:43 p.m.

Great movie!

Edit Delete

Create a review

Comment:

Add Review

Figure 1.10 – A specific movie page with its reviews

Many other functionalities will be developed and explained across the book. Now, let's see the architecture we will use to construct the *Movies Store* application.

Introducing the Django MVT architecture

There are various methodologies and approaches to design and code web applications. One approach involves consolidating all code into a single file to construct the entire web application. However, finding errors within such a file, often comprising thousands of lines of code, can be incredibly challenging. Alternatively, other strategies distribute code across different files and directories. Additionally, some approaches segment an application into multiple smaller applications dispersed across several servers, although managing the distribution of these servers presents its own set of challenges.

Organizing your code effectively presents challenges. This is why developers and computer scientists have created software architectural patterns. **Software architectural patterns** offer structural frameworks or layouts to address common software design issues. By leveraging these patterns, start-ups and inexperienced developers can avoid reinventing solutions for every new project. Various architectural patterns exist, including **Model-View-Controller** (**MVC**), **Model-View-Template** (**MVT**), layers, service-oriented, and microservices. Each pattern comes with its own set of pros and cons. Many frameworks, such as Django, adhere to specific patterns in constructing their applications.

In the case of Django, Django is based on the MVT pattern. This pattern is similar to MVC but with some differences in the responsibilities of each component:

- **Models**: The model represents the data structure. In Django, models are Python classes that define the structure of the data and how it interacts with the database. Models handle tasks such as querying a database, performing **CRUD** (**Create, Read, Update, Delete**) operations, and enforcing data validation. In the case of the *Movies Store* app, *Movie*, *Review*, *Order* and the other classes from our class diagram will be coded as Django models.

- **Views**: Views in Django are responsible for processing user requests and returning appropriate responses. Views typically receive HTTP requests from clients, fetch data from the database using models, and render templates to generate HTML responses. In Django, views are Python functions or classes that accept HTTP requests and return HTTP responses. In the case of the *Movies Store* app, we will create views and functions to handle the movies, accounts, and cart, among others.

- **Templates**: Templates are used to generate HTML dynamically. They contain the application's user interface and define how data from the views should be displayed to the users. In the case of the *Movies Store* app, we will create a template to allow users to log in, a template to list movies, and a template to display the shopping cart, among others.

The MVT pattern offers several benefits such as enhanced code separation, facilitated collaboration among multiple team members, simplified error identification, increased code reusability, and improved maintainability. *Figure 1.11* illustrates the software architecture of the *Movies Store*, which we will develop throughout this book. While it may seem overwhelming now, you will understand the intricacies of this architecture by the book's conclusion. We will delve deeper into the architecture in the final chapters.

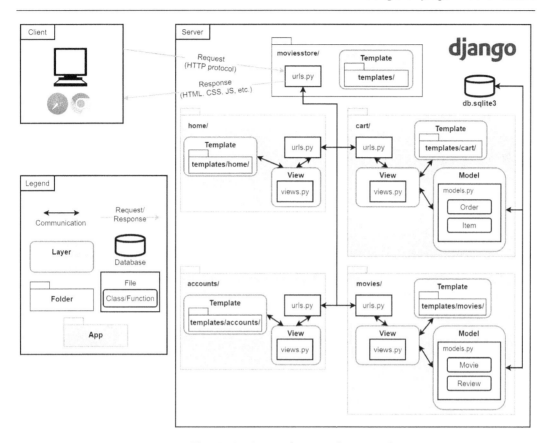

Figure 1.11 – The Movies Store software architecture diagram

Let's briefly analyze this architecture:

- Positioned on the left are the clients, which are the users of our application, who use browsers on mobile or desktop devices. These clients establish connections with the application via the **Hypertext Transfer Protocol** (**HTTP**), providing users with a means to interact with our web application.

- On the right side, we have the server, which hosts our application code.

- All client interactions first pass for a project-level URL file called urls.py. This file is located in the main project folder called moviesstore/. URLs will be explored in *Chapter 2*. This project folder also contains a templates/ folder in which we will design a reusable base template. Base templates will be explored in *Chapter 3*.

- The project-level URL file passes the interaction to an app-level URL file. For this project, we will design and implement four Django apps – *home*, *movies*, *cart*, and *accounts*. Django apps will be explored in *Chapter 2*.

- Each app-level URL file passes the interaction to a `views.py` file. Views will be explored in *Chapter 2*.

- Views communicate with models, if required, and pass information to the templates, which are finally delivered to the clients as HTML, CSS, and JS code. Templates will be explored in *Chapter 2*, and models will be explored in *Chapter 5*.

In *Figure 1.11*, the **Model**, **View**, and **Template** layers are highlighted in gray, representing the common architectural pattern used in Django, which will be utilized throughout this book. We have four models corresponding to the classes defined in our class diagram (as previously shown in *Figure 1.6*). The user model does not appear in this diagram because we will reuse a built-in Django user model.

Therefore, as mentioned earlier, there are different approaches to implementing web applications with Django. There are even different ways to implement a Django MVT architecture. In the following chapters, we will see the advantages of adopting an MVT architecture, as presented in *Figure 1.11*.

Summary

In this chapter, we learned how to install and use Python, `pip`, and Django. We also learned how to create a new Django project and run a Django local web server. Then, we explained the scope of the *Movies Store* project. We also illustrated the application data and its relationships through a class diagram. Additionally, we presented an architecture diagram that showed the main components and elements of the *Movies Store*. These diagrams will serve as a blueprint to codify the *Movies Store* project in the upcoming chapters.

In the next chapter, we will look inside the project folder that Django has created for us to understand it better, and we will create our first Django app.

2

Understanding the Project Structure and Creating Our First App

Django projects contain a predefined structure with some key folders and files. In this chapter, we will discuss the Django project structure and how some of those folders and files are used to configure our web applications. Furthermore, Django projects are composed of one or more apps. We will learn how to create a "home" app, composed of "home" and "about" sections, and how to register it inside our Django project.

In this chapter, we will cover the following topics:

- Understanding the project structure
- Creating our first app
- Creating a home page
- Creating an about page

With all of these topics completed, you will know how to create Django apps and web pages.

Technical requirements

In this chapter, we will be using Python 3.10+. Additionally, we will be using the **Visual Studio (VS) Code** editor in this book, which you can download from `https://code.visualstudio.com/`.

The code for this chapter is located at `https://github.com/PacktPublishing/Django-5-for-the-Impatient-Second-Edition/tree/main/Chapter02/moviesstore`.

The CiA video for this chapter can be found at `https://packt.link/rzU25`

Understanding the project structure

Let's look at the project files that were created for us in *Chapter 1*, in the *Creating and running a Django project* section. Open the `moviesstore` project folder in VS Code. You will see the elements shown in *Figure 2.1*:

Figure 2.1 – The MOVIESSTORE directory structure

Let's learn about each of these elements.

The moviesstore folder

As you can see in *Figure 2.1*, there is a folder with the same name as the folder we opened in VS Code originally – `moviesstore`. The `moviesstore` folder contains a set of files to configure the Django project. *Figure 2.2* shows the content of the `moviesstore` folder:

Figure 2.2 – The moviesstore folder content structure

Let's briefly look at all the elements under the `moviesstore` folder:

- `__pycache__`: This folder stores compiled bytecode when we generate our project. You can largely ignore this folder. Its purpose is to make your project start up a little faster by caching the compiled code, which can then be readily executed.

- `__init__.py`: This file indicates to Python that this directory should be considered a Python package. We can ignore this file.

- `asgi.py`: Django, being a web framework, needs a web server to operate. And since most web servers don't natively speak Python, we need an interface to make that communication happen. Django currently supports two interfaces – **Web Server Gateway Interface (WSGI)** and **Asynchronous Server Gateway Interface (ASGI)**. The `asgi.py` file contains an entry point for ASGI-compatible web servers to serve your project asynchronously.

- `settings.py`: The `settings.py` file is an important file that controls our project's settings. It contains several properties; let's analyze some of them:

  - `BASE_DIR`: Determines where on your machine the project is situated.

  - `SECRET_KEY`: This is a secret key for a particular Django project. It is used to provide cryptographic signing and should be set to a unique, unpredictable value. In a production environment, it should be replaced with a securely generated key.

  - `DEBUG`: Our site can run in debug mode or not. In debug mode, we get detailed information on errors, which is very useful when we develop our applications. For instance, if we try to run `http://localhost:8000/123` in the browser, we will see a **Page not found (404)** error (see *Figure 2.3*):

Figure 2.3 – Accessing an invalid application route

- INSTALLED_APPS: This setting specifies the list of all Django applications that are enabled for this project. Each string in the list represents the Python path to a Django application. By default, Django includes several built-in applications, such as admin, auth, contenttypes, and sessions. We will see later in this chapter how to create our own applications and how to include them in this configuration.

- MIDDLEWARE: Middleware in Django intercepts and manages the request and response processing flow. The listed middleware is provided by Django and handles various aspects of request/response processing, including security, session management, authentication, and more.

- ROOT_URLCONF: Specifies the Python path to the root URL configuration for the Django project.

- TEMPLATES: Defines the configuration for Django's template system. It includes information regarding the list of directories that the system should look in for template source files and other specific template settings.

- There are some other properties in settings.py, such as DATABASES, LANGUAGE_CODE, and TIME_ZONE, but we focused on the more important properties in the preceding list. We will later revisit this file and see how relevant it is when developing our site.

- urls.py: This file contains the URL declarations for this Django project. It could link specific URL paths to functions, classes, or other URL files to generate a response, or to render a page in response to a browser or URL request. We will later add paths to this file and better understand how it works.

- wsgi.py: This file contains an entry point for WSGI-compatible web servers to serve your project. By default, when we run the server with the python manage.py runserver command, it uses the WSGI configuration.

manage.py

The manage.py file seen in *Figure 2.1* and *Figure 2.2* is a crucial element that we will extensively use throughout this book. This file provides a command-line utility that lets you interact with a Django project and perform some administrative operations. For example, we earlier ran the following command in *Chapter 1*, in the *Creating and running a Django project* section:

```
python manage.py runserver
```

The purpose of the command was to start the local web server. We will later illustrate more administrative functions, such as one to create a new app – python manage.py startapp.

db.sqlite3

The db.sqlite3 file is the default SQLite database file that Django uses for development purposes. **SQLite** is a lightweight, serverless, and self-contained SQL database engine that doesn't require a separate server process to operate. It stores an entire database (including tables, indexes, and data) as a single file (into the db.sqlite3 file). We will not use this file for now; however, we will discuss it in *Chapter 5*.

We have learned about the Django project structure and some of its main elements. Now, let's create our first Django app.

Creating our first app

A **Django app** is a self-contained package of code that performs a specific functionality or serves a particular purpose within a Django project.

A single Django project can contain one or more apps that work together to power a web application. Django uses the concept of projects and apps to keep code clean and readable.

For example, on a movie review site such as *Rotten Tomatoes*, as shown in *Figure 2.4*, we can have an app for listing movies, an app for listing news, an app for payments, an app for user authentication, and so on:

Figure 2.4 – The Rotten Tomatoes website

Apps in Django are like pieces of a website. You can create an entire website with one single app, but it is useful to break it up into different apps, each representing a clear function.

Our *Movies Store* site will begin with one app. We will later add more as we progress. To add an app, in the Terminal, stop the server by pressing *Cmd+ C*. Navigate to the top moviesstore folder (the one that contains the manage.py file) and run the following in the Terminal:

For macOS, run the following command:

```
python3 manage.py startapp home
```

For Windows, run the following command:

```
python manage.py startapp home
```

A new folder, home, will be added to the project (see *Figure 2.5*). As we progress in the book, we will explain the files that are inside the folder.

Figure 2.5 – The MOVIESSTORE project structure containing the home app

Although our new home app exists in our Django project, Django doesn't recognize it till we explicitly add it. To do so, we need to specify it in settings.py. So, go to /moviesstore/settings.py, under INSTALLED_APPS, and you will see six built-in apps already there.

Add the app name, as highlighted in the following (this should be done whenever a new app is created):

```
...
INSTALLED_APPS = [
    'django.contrib.admin',
    'django.contrib.auth',
    'django.contrib.contenttypes',
    'django.contrib.sessions',
    'django.contrib.messages',
    'django.contrib.staticfiles',
    'home',
]
...
```

We have successfully created our first app and included it in our Django settings project. Now, we are going to create and serve two pages inside this app.

Creating a home page

Creating a simple page or section in Django usually involves three steps:

1. Configure a URL.

2. Define a view function or class.

3. Create a template.

Let's see how to apply those steps to create a simple "home" page that will display a "welcome" message to the final user.

Configuring an URL

Django URLs (Uniform Resource Locators) are patterns used to map incoming HTTP requests to the appropriate view functions or classes that handle those requests. They define the routing mechanism for your Django project, specifying which views should be called for different URLs.

There is a main URL configuration file located at /moviesstore/urls.py that currently has the following code:

```
...
from django.contrib import admin
from django.urls import path

urlpatterns = [
    path('admin/', admin.site.urls),
]
```

When a user types a URL (related to our Django application) in the browser, a request first passes through the `/moviesstore/urls.py` file, and it will try to match a `path` object in `urlpatterns` – for example, if a user enters `http://localhost:8000/admin` into the browser, the URL will match the `admin/` path. The server will then respond with the Django admin page (as shown in *Figure 2.6*), which we will explore later:

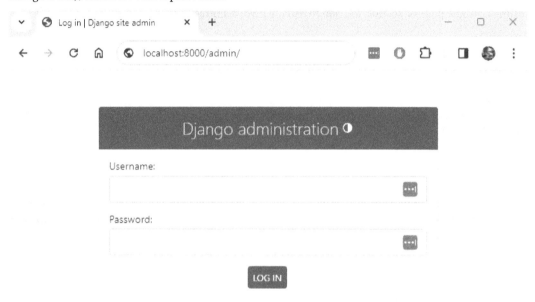

Figure 2.6 – The /admin route – the admin page

Conversely, if a user enters `localhost:8000/hello`, Django will return a `404 not found` page because there aren't any matching paths in the URL configuration file.

Creating a path for the home page

There are two ways to create a custom path for a page:

- Create the path in the project-level URL file (in `/moviesstore/urls.py`)
- Create the path in a `urls.py` file, defined at the app level (in `/home/urls.py`).

We will use the second option in this book, since it allows us to keep our URLs separate and organized.

In `/home/`, create a new file called `urls.py`. This file will contain the path relating to the URLs of the home app. For now, fill it in with the following:

```
from django.urls import path
from . import views
```

```
urlpatterns = [
    path('', views.index, name='home.index'),
]
```

Let's explain the previous code:

- We import the `path` function, which is used to define URL patterns in Django.

- We import the `views` file. In the next section, we will implement an `index` function inside the `views` file. That function will render a template that contains a "welcome" message.

- We define the `urlpatterns` for the home app. In this case, inside the `urlpatterns` list, we add a new path object with three arguments:

  - The first argument, `' '`, represents the URL pattern itself. In this case, it's an empty string, indicating the root URL. This means that when the root URL of the application is accessed (`localhost:8000/`), it will match this path.

  - The second argument, `views.index`, refers to the view function that will handle the HTTP request. Here, `views.index` indicates that the `index` function in the `views` file is responsible for processing the request.

  - The third argument, `name='home.index'`, is the name of the URL pattern. This name is used to uniquely identify the URL pattern and can be referenced in other parts of the Django project, such as templates or other URL patterns.

Now, let's proceed to define the `views.index` function code.

Defining a view function

Django views are Python functions or classes that receive web requests and return web responses. They contain the logic to process HTTP requests and generate appropriate HTTP responses, typically in the form of HTML content to be rendered in the user's web browser.

Our home app already includes a `views.py` file; let's take advantage of it and make a simple modification. In `/home/views.py`, add the following in **bold**:

```
from django.shortcuts import render

def index(request):
    return render(request, 'home/index.html')
```

Let's explain the previous code:

- By default, the `views` file imports the `render` function, which is used to render templates and return an HTTP response with the rendered content.

- We define an `index` function. This function takes one parameter, `request`, which represents the HTTP request received by the server.

- Finally, the `index` function returns a rendered template. The `render` function takes the `request` as the first argument, and the second argument (`'home/index.html'`) represents the path to the template file to be rendered. In the next section, we will create that template.

We have now connected the `''` path with the proper `views.index` function, but we are missing the connection between the `views.index` function and the `'home/index.html'` template. So, let's implement the template.

Creating a template

Django templates are text files containing HTML, along with **Django template language** (DTL) syntax, which describes the structure of a web page. Django templates allow you to dynamically generate HTML content by inserting variables, loops, conditionals, and other logic into the HTML markup.

Our "home" app doesn't include a location to store templates, so let's create it. In `/home/`, create a `templates` folder. Then, in `/home/templates/`, create a `home` folder.

Now, in `/home/templates/home/`, create a new file, `index.html`. This will be the full HTML page for the "home" page. For now, fill it in with the following:

```html
<!DOCTYPE html>
<html>
<head>
  <title>Home page</title>
</head>
<body>
  <h1>Welcome to the Home Page</h1>
</body>
</html>
```

This file contains a simple HTML code with a "welcome" message.

> **Note**
>
> We suggest storing your app templates under the next directory structure – `app_name/templates/app_name/my_template.html`. Sometimes, different apps can contain templates with the same name, which could lead to potential name conflicts in template resolution. By using the previous strategy, you can define templates with the same name in different Django apps without any potential name conflict.

We have completed the connection between the URL, view function, and template. However, Django doesn't know how to use our /home/urls.py file. So, let's connect this file to our main URL configuration file, and then we will have completed the puzzle.

Connecting a project-level URL file with an app-level URL file

In /moviesstore/urls.py, add the following in **bold**:

```
...
from django.contrib import admin
from django.urls import path, include

urlpatterns = [
    path('admin/', admin.site.urls),
    path('', include('home.urls')),
]
```

Let's explain the previous code:

- We modify the code to also import the include function, which is used to include URLs from other URL configuration files.

- We add a new path object to the urlpatterns list. The empty string, ' ', represents the base URL to include the URLs from the home.urls file.

Now, save those files, run the server, and go back to http://localhost:8000; you should see the home page displayed (*Figure 2.7*):

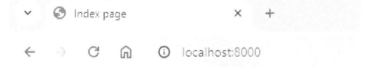

Welcome to the Home Page

Figure 2.7 – The home page

> **Note**
>
> When we make changes to a file and save it, Django observes the file changes and reloads the server with them. Therefore, we don't have to manually restart the server each time there is a code change.

Now that we have our "home" page up and running, let's repeat the process to create the "about" page.

Creating an about page

Now that we learned how to create a simple page, let's repeat the process to create the about page. We will follow these three steps:

1. Configure the about URL.

2. Define the about function.

3. Create the about template.

Let's start.

Configuring the about URL

In /home/urls.py, add the following path in **bold**:

```
from django.urls import path
from . import views

urlpatterns = [
    path('', views.index, name='home.index'),
    path('about', views.about, name='home.about'),
]
```

So, if a URL matches the /about path, it will execute the about function defined in the views file.

Defining about function

In /home/views.py, add the following in **bold**:

```
from django.shortcuts import render

def index(request):
    return render(request, 'home/index.html')

def about(request):
    return render(request, 'home/about.html')
```

The about function is similar to the index function. This function renders the 'home/about.html' template, which will be implemented next.

Creating about template

Now, in `/home/templates/home/`, create a new file, `about.html`. This file contains the HTML for the about page. For now, fill it in with the following:

```html
<!DOCTYPE html>
<html>
<head>
  <title>About page</title>
</head>
<body>
  <h1>Welcome to the About Page</h1>
</body>
</html>
```

Save the files, and when you navigate to `localhost:8000/about`, it will show the about page (*Figure 2.8*):

Welcome to the About Page

Figure 2.8 – The about page

> **Note**
>
> When we executed the command to create the home app, some folders and files were automatically created for us. For the home app, we won't use many of them. So, you can optionally delete the following folders and files to keep your application clean and simple – `migrations/`, `admin.py`, `models.py`, and `tests.py`.

We quickly created our second page, "about." Now, we hope you have a better understanding of how URLs, views, and templates connect.

Summary

In this chapter, we discussed the Django project structure. We analyzed some of the most important project folders, files, and their functionalities. We saw how a web project can be composed of several applications, and we learned how to create a Django app. We also learned how URLs, views, and templates connect to create web pages. We created a couple of pages and loaded them into our local web server. In the next chapter, we will see how to improve the look and feel of our Django applications by using base templates and a CSS framework.

3

Designing a Base Template

Django projects can consist of dozens or hundreds of template files. Sometimes, these files can contain duplicated HTML and CSS code, which affects the project's maintainability. In this chapter, we introduce the concept of **base templates** and how they can be used to reduce duplicated template code. We will also improve the look and feel of our application by designing a base template that includes a header and a footer, as well as links to different pages.

In this chapter, we will be covering the following topics:

- Creating a base template with Bootstrap
- Updating the **Home** page to use the base template
- Updating the **About** page to use the base template
- Adding a header section
- Adding a footer section

In the end, you will learn the importance of base templates and how they can be used to reduce duplicated code and improve the look and feel of your web applications.

Technical requirements

In this chapter, we will be using **Python 3.10+**. Additionally, we will be using the **VS Code** editor in this book, which you can download from `https://code.visualstudio.com/`.

The code for this chapter is located at `https://github.com/PacktPublishing/Django-5-for-the-Impatient-Second-Edition/tree/main/Chapter03/moviesstore`.

The CiA video for this chapter can be found at `https://packt.link/psU29`

Creating a base template with Bootstrap

We currently have two templates (`index.html` and `about.html`) that duplicate the structure of the site and some HTML tags. Currently, it doesn't seem like a serious issue. However, once the application starts growing, we will have a lot of duplicated HTML code spread over dozens of template files. To avoid this issue, we will create a base template that contains the main structure of the site over a single file, and the other templates will extend this base template.

Introducing Bootstrap

Bootstrap is the most popular CSS framework for developing responsive and mobile-first websites (see *Figure 3.1*). Bootstrap provides a set of HTML, CSS, and JavaScript components and utilities that developers can use to build modern user interfaces quickly. For Django projects, a developer can design the user interface from scratch if they want to. However, as this book is not about user interfaces, we will take advantage of CSS frameworks (such as Bootstrap) and use some of their elements and examples to create something that looks professional. You can find out more about Bootstrap at `https://getbootstrap.com/`.

Figure 3.1 – The Bootstrap site

Introducing Django template language (DTL)

We will build the base template as a combination of Bootstrap, HTML, CSS, JavaScript, and **DTL**.

DTL is a templating language used within the Django web framework for building dynamic web pages (`https://docs.djangoproject.com/en/5.0/topics/templates/`). It is designed to separate the presentation layer from the business logic of an application, promoting clean and maintainable code.

Some Django template language key features include the following:

- **Double curly braces**: Variables, expressions, and template tags are enclosed within double curly braces. For example, `{{ variable }}`.

- **Template tags**: Control structures and logic are defined within template tags, which are enclosed within `{% %}`. Template tags allow for loops, conditionals, and other control flow statements. For example, `{% if condition %} ... {% endif %}`.

- **Comments**: Comments in DTL are enclosed within `{# #}` and are not rendered in the final output HTML.

- **Template inheritance**: Django templates support inheritance, allowing for the creation of base templates that define the overall structure and layout of a page, with child templates inheriting and overriding specific blocks or sections.

Creating a base template

The base template will serve as a "global" template (which will be used across all pages and apps). So, we will add it to our main project folder. In the `moviesstore/` folder (the directory that contains the `settings.py` file), create a folder called `templates`. In that folder, create a file called `base.html`. For now, fill it in with the following:

```html
<!DOCTYPE html>
<html>
  <head>
    <title>{{ template_data.title }}</title>
    <link href=
      "https://cdn.jsdelivr.net/npm/bootstrap@5.3.3/
      dist/css/bootstrap.min.css" rel="stylesheet"
      crossorigin="anonymous">
    <link rel=
      "stylesheet"  href="https://cdnjs.cloudflare.com/
      ajax/libs/font-awesome/6.1.1/css/all.min.css">
    <link href=
      "https://fonts.googleapis.com/
      css2?family=Poppins:wght@300;400;500;600;700&display=
      swap" rel="stylesheet">
    <script src=
      "https://cdn.jsdelivr.net/npm/bootstrap@5.3.3/dist/
```

```
       js/bootstrap.bundle.min.js"crossorigin="anonymous">
     </script>
     <meta name="viewport" content="width=device-width,
       initial-scale=1" />
   </head>

   <body>
     <!-- Header -->
     <!-- Header -->

     <div>
       {% block content %}
       {% endblock content %}
     </div>

     <!-- Footer -->
     <!-- Footer -->
   </body>
 </html>
```

The previous file contains a base HTML structure for our site. Let's review some important aspects of the previous code:

- The head tag contains the title tag, which uses DTL double curly braces to display the information of a variable ({{ template_data.title }}). Later, we will see how to pass that variable from views to this template. It also contains some links and a script to include Bootstrap and some fonts for our site. We take some of those links from this site: https://getbootstrap.com/docs/5.3/getting-started/introduction/#cdn-links.

- The body tag contains an HTML comment indicating the location of the header (we will later include the header in that position) and div, which includes a couple of DTL template tags. {% block %} and {% endblock %} are template tags used for template inheritance. This is a template tag that defines a block named content. Blocks are placeholders in the template that can be overridden by child templates. The content within this block will be replaced by the content defined in a child template that extends this template (we will see it later in action). It also contains an HTML comment indicating the location of the footer (we will later include the footer in that position).

Registering the base template

Finally, we need to register the `moviesstore/templates` folder in our application settings. We need to import the `os` module and include the new template path in our `/moviesstore/settings.py` file. In `/moviesstore/settings.py`, add the following in bold:

```
...
import os
from pathlib import Path
...
ROOT_URLCONF = 'moviesstore.urls'

TEMPLATES = [
    {
        'BACKEND': 'django.template.backends.django.
                   DjangoTemplates',
        'DIRS': [os.path.join(BASE_DIR,
                              'moviesstore/templates')],
        'APP_DIRS': True,
        ...
```

Now that we have defined our base template structure, let's update the **Home** and **About** pages to extend this template.

Updating the home page to use the base template

The new home page will extend the base template; it will include a background with an image and it will include custom CSS. Let's create the new home page.

Creating the new index template

In `/home/templates/home/index.html`, replace the entire template code with the following:

```
{% extends 'base.html' %}
{% block content %}
<header class="masthead bg-index text-white text-center
          py-4">
  <div class="container d-flex align-items-center flex-
  column pt-2">
    <h2>Movies Store</h2>
    <p>Your Ticket to Unlimited Entertainment!</p>
  </div>
</header>
<div class="p-3">
```

```
<div class="container">
  <div class="row mt-3">
    <div class="col mx-auto text-center mb-3">
      <h4>Welcome to the best movie store!!</h4>
    </div>
  </div>
</div>
</div>
{% endblock content %}
```

Let's explain the previous code:

- The new index.html file now extends the base.html template.

- The code that is inside {% block content %} {% endblock content %} will be injected inside div of the base.html template file. This code defines a couple of messages and uses some custom CSS classes that will be defined next.

Creating a custom CSS file

In the moviesstore/ folder (the directory that contains the settings.py file), create a folder called static. In that folder, create a subfolder called css. Then, in moviesstore/static/css/ create a file called style.css. For now, fill it in with the following:

```
.bg-index{
  background: url("/static/img/background.jpg") no-repeat
    fixed;
  background-size: 100% auto;
}
```

The previous code defines a CSS class called bg-index, which will be used to display an image as a background on the home page.

Storing an image

Let's also include the background.jpg image in our project. In moviesstore/static, create a folder called img. Then, in moviesstore/static/img/, download and store the background.jpg image from this link: https://github.com/PacktPublishing/Django-5-for-the-Impatient-Second-Edition/blob/main/Chapter03/moviesstore/moviesstore/static/img/background.jpg (as shown in *Figure 3.2*).

Figure 3.2 – Including a background image under the project structure

Serving the static files

We have defined a couple of static files, a CSS file, and a JPG file. To be able to use them or display them in our application, we need to register the folder that contains them. Add the following code in **bold** at the end of the /moviesstore/settings.py file:

```
...

DEFAULT_AUTO_FIELD = 'django.db.models.BigAutoField'

STATICFILES_DIRS = [
    BASE_DIR / 'moviesstore/static/',
]
```

Updating the base template to use the custom CSS and load static files

We also need to update the base template to link the custom CSS we previously created, and we need to use a custom DTL tag to load the static files. In `/moviesstore/templates/base.html`, add the following in bold:

```
<!DOCTYPE html>
<html>
  {% load static %}
  <head>
    <title>{{ template_data.title }}</title>
    ...
    <link rel="stylesheet" type="text/css"
          href="{% static 'css/style.css' %}">
    <meta name="viewport"
          content="width=device-width, initial-scale=1" />
  </head>
  ...
```

In the previous code, the `load static` template tag is used to load the static files in the `base.html` template. Once we have used this tag, we can use the `static` template tag to refer to specific static files to be loaded. It will search for the static files based on the `STATICFILES_DIRS` folder location.

Now, save those files, run the server, and go back to `http://localhost:8000`; you should see the new home page displayed (*Figure 3.3*). Check that the tab title doesn't appear, as we need to send the `template_data.title` variable from the view function to the template (which is carried out next).

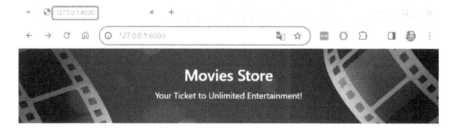

Welcome to the best movie store!!

Figure 3.3 – The new home page with the missing tab title

> **Note**
>
> If you have problems loading the background image, we recommend you stop the server and run it again or clear the browser cache. Also, try to access the image file directly from the browser to check whether the image was loaded properly (`http://localhost:8000/static/img/background.jpg`).

Updating the views index function

Finally, let's pass the title from the view function to the templates. In `/home/views.py`, add the following in bold:

```python
from django.shortcuts import render

def index(request):
    template_data = {}
    template_data['title'] = 'Movies Store'
    return render(request, 'home/index.html', {
        'template_data': template_data})

def about(request):
    return render(request, 'home/about.html')
```

Let's explain the previous code:

- We create a Python dictionary called `template_data`. We will always use this dictionary when we need to pass information from view functions to templates.

- We add a key called `title` to the `template_data` dictionary. `title` will be used to define the browser tab title. Remember that `template_data.title` is used in the `base.html` template.

- We modify the `render` function to pass a third argument. This time we pass the `template_data` variable, which will be available across the `home/index.html` template or the templates that it extends.

Figure 3.4 displays the updated **Home** page with the proper browser tab title.

Welcome to the best movie store!!

Figure 3.4 – New home page with proper browser tab title

Updating the About page to use the base template

The new **About** page will also extend the base template, and it will include a dummy text about the page and an image.

Creating the new About template

In /home/templates/home/about.html, replace the entire template code with the following:

```
{% extends 'base.html' %}
{% block content %}
{% load static %}
<div class="p-3">
  <div class="container">
    <div class="row mt-3">
      <div class="col-md-6 mx-auto mb-3">
        <h2>About</h2>
        <hr />
        <p>
          At Movies Store, we offer a vast digital library
          that spans across genres, ensuring there's
          something for every movie lover. Browse our
          extensive collection of films, including the
          latest releases, timeless classics, and hidden
          gems. With just a few clicks, you can rent or
          purchase your favorite titles and instantly
          stream them in high-definition quality.
        </p>
        <p>
          Discover the convenience of our digital platform,
```

```
            where you have the flexibility to watch movies
            on your preferred device, whether it's a smart
            TV, tablet, or smartphone. With our intuitive
            search and recommendation features, finding your
            next movie night pick has never been easier.
          </p>
        </div>
        <div class="col-md-6 mx-auto mb-3 text-center">
          <img src="{% static 'img/about.jpg' %}"
               class="max-width-100"
               alt="about" />
        </div>
      </div>
    </div>
  </div>
{% endblock content %}
```

Let's explain the previous code:

- The new about.html file now extends the base.html template.

- We use {% block content %} {% endblock content %} to inject the proper HTML code inside div of the base.html template file. This code defines a paragraph about the page and displays an image that will be stored next.

- We also use the {% load static %} tag since this template loads a custom image by using the static template tag.

Storing the about.jpg image

Let's also include the about.jpg image in our project. In moviesstore/static/img/, download and store the about.jpg from this link: https://github.com/PacktPublishing/Django-5-for-the-Impatient-Second-Edition/blob/main/Chapter03/moviesstore/moviesstore/static/img/about.jpg.

Updating the views about function

Finally, let's pass the title from the view about function to the templates. In /home/views.py, add the following in bold:

```
from django.shortcuts import render
...
def about(request):
    template_data = {}
    template_data['title'] = 'About'
```

```
return render(request,
              'home/about.html',
              {'template_data': template_data})
```

Similar to the `index` function, we define the `template_data` dictionary and create the proper `title` key with its value. Then, we pass the `template_data` variable to the templates.

Now, save those files, run the server, and go to `http://localhost:8000/about`; you should see the new **About** page displayed (*Figure 3.5*).

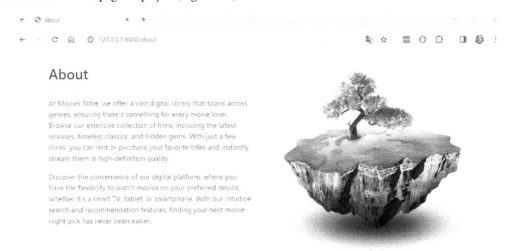

Figure 3.5 – New About page

Now that we have updated the home and **About** pages, let's improve the base template by adding a header section that includes the website menu options.

Adding a header section

To complete the `base.html` template, we need to include a header section and a footer section. Let's start with the header.

Updating the base template

In `/moviesstore/templates/base.html`, add the following in bold:

```
...
<body>
  <!-- Header -->
  <nav class="p-3 navbar navbar-dark bg-dark
      navbar-expand-lg">
```

```
      <div class="container">
        <a class="navbar-brand"
           href="{% url 'home.index' %}">
          <img src="{% static 'img/logo.png' %}" alt="logo"
               height="40" />
        </a>
        <button class="navbar-toggler" type="button"
                data-bs-toggle="collapse"
                data-bs-target="#navbarNavAltMarkup"
                aria-controls="navbarNavAltMarkup"
                aria-expanded="false"
                aria-label="Toggle navigation">
          <span class="navbar-toggler-icon"></span>
        </button>
        <div class="collapse navbar-collapse"
             id="navbarNavAltMarkup">
          <div class="navbar-nav ms-auto navbar-ml">
            <a class="nav-link"
               href="{% url 'home.about' %}">About</a>
          </div>
        </div>
      </div>
    </nav>
    <!-- Header -->
    …
```

We included a responsive `navbar` between the `Header` HTML comments. This responsive navbar includes a `logo.png` file that links to the `home.index` URL, and includes an `About` text that links to the `home.about` URL. Check that we used the `url` template tag, as this tag links to the specified URL pattern name.

> **Note**
>
> The construction of the previous header section is inspired by the Bootstrap `navbar` component. You can take a look at this component and its available options at this link: `https://getbootstrap.com/docs/5.3/components/navbar/`.

Storing the logo image

Let's include the `logo.png` image in our project. In `moviesstore/static/img/`, download and store the `logo.png` image from this link: `https://github.com/PacktPublishing/Django-5-for-the-Impatient-Second-Edition/blob/main/Chapter03/moviesstore/moviesstore/static/img/logo.png`.

Updating the style.css

Finally, let's include a couple of CSS classes in our custom CSS file. In `/moviesstore/static/css/style.css`, add the following in bold at the end of the file:

```
...

.navbar a.nav-link {
  color: #FFFEF6 !important;
}

.bg-dark {
  background-color: #2E2E2E !important;
}
```

Now, save those files, run the server, and go to `http://localhost:8000/`; you should see the home page with the new header section (*Figure 3.6*).

Figure 3.6 – The home page with the header section

This new header section is responsive. If you reduce the browser window width, you will see a responsive navbar, thanks to the use of different Bootstrap classes (see *Figure 3.7*).

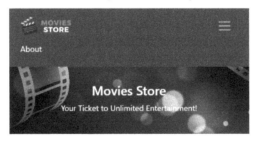

Figure 3.7 – Home page with a responsive navbar

The base template now includes a proper header section. Let's finalize this template by adding a footer section.

Adding a footer section

Let's complete the website structure with the inclusion of a footer.

Updating the base template

In /moviesstore/templates/base.html, add the following in bold:

```
...
<!-- Footer -->
<section class="p-3 ms-footer d-none d-md-block">
  <div class="container">
    <div class="row mt-3 text-white">
      <div class="col-md-6 col-lg-6 col-xl-6
          mx-auto mb-4">
        <b>MOVIES STORE</b>
        <hr />
        <p>
          Welcome to Movies Store, your premier online
          destination for a cinematic adventure like no
          other! Dive into the world of movies from the
          comfort of your own home with our user-
          friendly and immersive online movie store.
        </p>
      </div>
      <div class="col-md-3 col-lg-3 col-xl-3
          mx-auto mb-4">
        <b>LINKS</b>
        <hr />
        <p><a class="nav-link"
          href="{% url 'home.about' %}">
          About
        </a></p>
      </div>
      <div class="col-md-3 col-lg-3 col-xl-3 mx-auto
          mb-4">
        <b>CONTACT</b>
        <hr />
        <p><i class="fas fa-home me-2"></i>
          150-2345 Tokyo-to, Japan
```

```
            </p>
            <p><i class="fas fa-envelope me-2"></i>
              info@moviesstore.com
            </p>
            <p><i class="fas fa-phone me-2"></i>
              +81 03-3333-3333
            </p>
          </div>
        </div>
      </div>
    </section>
    <section class="p-3 ms-footer-bottom bg-dark">
      <div class="container d-flex
          justify-content-between">
        <div class="me-5 text-white">
          <span>DESIGNED BY
            <a href="https://www.x.com/danielgarax"
            target="_blank">DANIEL CORREA</a> &
            <a href="https://www.x.com/greglim81"
            target="_blank">GREG LIM</a>
          </span>
        </div>
        <div class="text-white">
          <span>© Copyright - 2024</span>
        </div>
      </div>
    </section>
    <!-- Footer -->
    ...
```

We included a footer section with information on the website, some links, and the book's author names and links to their X accounts.

Updating the style.css

Finally, let's include some custom CSS classes. In /moviesstore/static/css/style.css, add the following in bold at the end of the file:

```
...
.ms-footer {
  background-color: #202020;
}

.ms-footer p {
```

```
  color: #7F7F7F;
  font-size: 13px;
}

.ms-footer a:hover {
  color: #6ab43e;
  text-decoration: none;
}

.ms-footer-bottom span{
  font-size: 13px;
  line-height: 38px;
}

.ms-footer-bottom a {
  color: #6ab43e;
  text-decoration: none;
}

.ms-footer-bottom a:hover {
  color: #fff;
}
```

Now, save those files, run the server, and go to `http://localhost:8000/`; you should see the home page with the new footer section (*Figure 3.8*).

Welcome to the best movie store!!

Figure 3.8 – The home page with the footer section

You can also click the **About** link, and you will see the **About** page with the same website structure.

Summary

In this chapter, we learned how to create base templates that reduce duplicated code. We improved our application interface with the inclusion of a header and footer, and we learned how to manage static files. We redesigned the home and **About** pages to extend the base template and created proper links to those pages. In the next chapter, we'll learn how to start managing movies.

4

Creating a Movies App with Dummy Data

Currently, our project contains a single application with a couple of sections that display static information. Web applications are more complex. In this chapter, we will learn how to develop more complex applications, such as the movies app. The movies app will serve to list movies and enable users to click on them to display their data on a separate page. For now, we will use dummy data to simulate the movie data.

In this chapter, we will be covering the following topics:

- Creating the movies app
- Listing movies with dummy data
- Listing individual movies
- Adding a link in the base template

By the end, we will know how to create more complex Django apps and how to manage information inside those apps.

Technical requirements

In this chapter, we will be using Python 3.10+. Additionally, we will be using the **VS Code** editor in this book, which you can download from https://code.visualstudio.com/.

The code for this chapter is located at https://github.com/PacktPublishing/Django-5-for-the-Impatient-Second-Edition/tree/main/Chapter04/moviesstore.

The CiA video for this chapter can be found at https://packt.link/WmJR1

Creating the movies app

Currently, we have a home app that contains the logic to navigate between the **Home** and **About** pages. Now, we are going to start designing and implementing the movies logic. We prefer to separate this logic from the home app. So, let's create a new Django app. We will follow the next steps: (i) creating the movies app, (ii) adding the movies app to settings, and (iii) including the movies URL file in the project-level URL file.

Creating the movies app

Navigate to the top `moviesstore` folder (the one that contains the `manage.py` file) and run the following in the Terminal:

For macOS, run the following command:

```
python3 manage.py startapp movies
```

For Windows, run the following command:

```
python manage.py startapp movies
```

Figure 4.1 shows the new project structure. Verify that it matches your current folder structure.

Figure 4.1 – The MOVIESSTORE project structure containing the movies app

Adding the movies app to settings

Remember that for each newly created app, we must register it in the settings.py file. In /moviesstore/settings.py, under INSTALLED_APPS, add the following in bold:

```
...
INSTALLED_APPS = [
    'django.contrib.admin',
    'django.contrib.auth',
    'django.contrib.contenttypes',
    'django.contrib.sessions',
    'django.contrib.messages',
    'django.contrib.staticfiles',
    'home',
    'movies',
]
...
```

Including the movies URL file in the project-level URL file

In /moviesstore/urls.py, add the following in bold:

```
...
from django.contrib import admin
from django.urls import path, include

urlpatterns = [
    path('admin/', admin.site.urls),
    path('', include('home.urls')),
    path('movies/', include('movies.urls')),
]
```

Similar to the inclusion of the home.urls file, we include the movies.urls file, which will contain the URLs with respect to the movies app. All the URLs defined in the movies.urls file will contain a movies/ prefix (as defined in the previous path). We will create the movies.urls file later.

Now that we have created and included the movies app, we are ready to code the functionalities of this app. Let's start by listing movies.

Listing movies with dummy data

Listing movies involves a series of steps similar to those followed when we implemented the **Home** and **About** pages. We will follow the next steps: (i) configuring the movies URL, (ii) defining the views index function, and (iii) creating a movies index template.

Configuring the movies URL

In /movies/, create a new file called urls.py. This file will contain the path regarding the URLs of the movies app. For now, fill it in with the following:

```
from django.urls import path
from . import views

urlpatterns = [
    path('', views.index, name='movies.index'),
]
```

We defined a ' ' path, but remember that the project-level URLs file defined a /movies prefix for this file. So, if a URL matches the /movies path, it will execute the index function defined in the views file. We will implement the index function next.

Defining the views index function

In /movies/views.py, add the following in bold:

```
from django.shortcuts import render

movies = [
    {
        'id': 1, 'name': 'Inception', 'price': 12,
        'description': 'A mind-bending heist thriller.'
    },
    {
        'id': 2, 'name': 'Avatar', 'price': 13,
        'description': 'A journey to a distant world and
        the battle for resources.'
    },
    {
        'id': 3, 'name': 'The Dark Knight', 'price': 14,
        'description': 'Gothams vigilante faces the Joker.'
    },
    {
        'id': 4, 'name': 'Titanic', 'price': 11,
        'description': 'A love story set against the
        backdrop of the sinking Titanic.',
    },
]
```

```
def index(request):
    template_data = {}
    template_data['title'] = 'Movies'
    template_data['movies'] = movies
    return render(request, 'movies/index.html',
                  {'template_data': template_data})
```

Let's explain the previous code:

- We defined a variable called `movies`. This variable is a list of dictionaries, where each dictionary represents information about a particular movie. For example, at index 0, we have the movie with `id=1` (the `Inception` movie). We have four dummy movies. We will retrieve movie data from a SQLite database in upcoming chapters.

- We also have an `index` function. This function will render the `movies/index.html` template, but first, it passes a page title and the complete list of movies to that template.

Creating a movies index template

In `/movies/`, create a `templates` folder. Then, in `/movies/templates/`, create a `movies` folder.

Now, in `/movies/templates/movies/`, create a new file, `index.html`. For now, fill it in with the following:

```
{% extends 'base.html' %}
{% block content %}
{% load static %}
<div class="p-3">
  <div class="container">
    <div class="row mt-3">
      <div class="col mx-auto mb-3">
        <h2>List of Movies</h2>
        <hr />
      </div>
    </div>
    <div class="row">
      {% for movie in template_data.movies %}
      <div class="col-md-4 col-lg-3 mb-2">
        <div class="p-2 card align-items-center pt-4">
          <img src="{% static 'img/about.jpg' %}"
            class="card-img-top rounded">
          <div class="card-body text-center">
            {{ movie.name }}
          </div>
        </div>
      </div>
```

```
        </div>
        {% endfor %}
      </div>
    </div>
  </div>
{% endblock content %}
```

Let's explain the previous code:

- We extend the `base.html` template.

- We define a heading element with the text `List of Movies`.

- We use the DTL `for` template tag to iterate through each movie, and we display the movie name. For now, we are showing a default image for all movies; we will upload and display proper images for each movie in upcoming chapters.

> **Note**
>
> We used the Bootstrap card component as a base to design the way movies are displayed. You can find more information here: `https://getbootstrap.com/docs/5.3/components/card/`.

Now, save those files, run the server, and go to `http://localhost:8000/movies`; you should see the new **List of Movies** page (*Figure 4.2*).

Figure 4.2 – The List of Movies page

We are able to see the information of all movies together. Now, let's implement a functionality to list individual movies.

Listing individual movies

To list individual movies, we will follow these steps: (i) configuring individual movies URLs, (ii) defining the views show function, (iii) creating a movies show template, and (iv) adding individual movie links on the movies page.

Configuring individual movies URLs

In /movies/urls.py, add the next path in bold:

```
from django.urls import path
from . import views

urlpatterns = [
    path('', views.index, name='movies.index'),
    path('<int:id>/', views.show, name='movies.show'),
]
```

This path is a little different from the previously defined paths. The <int:id> part indicates that this path expects an integer value to be passed from the URL and that the integer value will be associated with a variable named id, which will be used to identify which movie data to show. For example, if we access movies/1, the application will display the data of the movie with id=1. Finally, that path will execute the show function defined in the views file. You can learn more about Django URLs here: https://docs.djangoproject.com/en/5.0/topics/http/urls/.

Defining the views show function

In /movies/views.py, add the following in bold at the end of the file:

```
...

def show(request, id):
    movie = movies[id - 1]

    template_data = {}
    template_data['title'] = movie['name']
    template_data['movie'] = movie
    return render(request, 'movies/show.html',
                  {'template_data': template_data})
```

Let's explain the previous code:

- We define the show function. This function takes two parameters: request and id (id is collected from the URL).

- Then, we extract the movie data with that ID. We subtract one unit since we stored the movie with `id=1` in the movies list index `0`, the movie with `id=2` in the movies list index `1`, and so on.

- Finally, we pass the movie name and the individual movie to the `movies/show.html` template.

Creating a movies show template

In `/movies/templates/movies/`, create a new file, `show.html`. For now, fill it in with the following:

```
{% extends 'base.html' %}
{% block content %}
{% load static %}
<div class="p-3">
  <div class="container">
    <div class="row mt-3">
      <div class="col-md-6 mx-auto mb-3">
        <h2>{{ template_data.movie.name }}</h2>
        <hr />
        <p><b>Description:</b> {{
          template_data.movie.description }}</p>
        <p><b>Price:</b> ${{
          template_data.movie.price }}</p>
      </div>
      <div class="col-md-6 mx-auto mb-3 text-center">
        <img src="{% static 'img/about.jpg' %}"
          class="rounded" />
      </div>
    </div>
  </div>
</div>
{% endblock content %}
```

The previous code displays the individual movie information.

Adding individual movie links on the movies page

In `/movies/templates/movies/index.html`, add the following in bold:

```
...
{% for movie in template_data.movies %}
<div class="col-md-4 col-lg-3 mb-2">
  <div class="p-2 card align-items-center pt-4">
    <img src="{% static 'img/about.jpg' %}"
```

```
        class="card-img-top rounded">
    <div class="card-body text-center">
        <a href="{% url 'movies.show' id=movie.id %}"
            class="btn bg-dark text-white">
            {{ movie.name }}
        </a>
    </div>
    </div>
    </div>
    {% endfor %}
    ...
```

We added a link from each movie name to each individual movie page. We used the `url` template tag to link to the specified URL pattern name (`movie.show`). But we also specified a parameter to be passed to the URL (`id=movie.id`). In this case, it's setting the `id` parameter to the `id` attribute of the `movie` object. This is useful for URLs that require dynamic parts, such as details for a specific movie.

Now, save those files, run the server, and go to `http://localhost:8000/movies`. You will see that each movie name has become a button that can be clicked. Click on a movie name, and you will be redirected to the individual movie page (*Figure 4.3*).

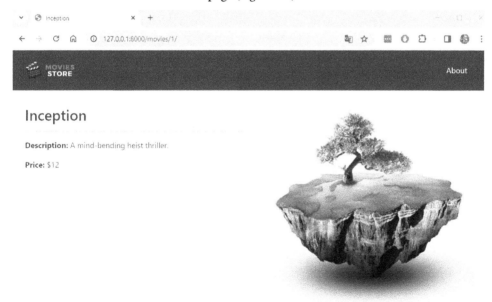

Figure 4.3 – Individual movie page

We can list all movies and navigate to individual movies; however, we haven't added a link to the movies section. Let's implement this link in the next section.

Adding a link in the base template

Finally, let's add the movies link in the base template. In /moviesstore/templates/base.html, in the header section, add the following in bold:

```
...
<div class="collapse navbar-collapse"
            id="navbarNavAltMarkup">
  <div class="navbar-nav ms-auto navbar-ml">
    <a class="nav-link" href=
      "{% url 'home.about' %}">About</a>
    <a class="nav-link" href=
      "{% url 'movies.index' %}">Movies</a>
  </div>
</div>
...
```

Now, save those files, run the server, and go to http://localhost:8000/movies. You will see the new **Movies** menu option in the header (*Figure 4.4*).

List of Movies

Figure 4.4 – Movies page updated

Summary

In this chapter, we recapped how to create a Django app. We created a movies app that allows listing movies and individual movies. We learned how to pass information through the URLs, how to create dummy data, how to use the `for` template tag, and how to link different pages. We hope this serves as a solid foundation to move on to the next part of our project, where we will go through more advanced topics such as models to make our site database-driven.

5

Working with Models

Storing data in a database is a common practice in most web applications. In a Django project, it involves working with Django models. In this chapter, we will create a database model (for example, a movie) and Django will turn this model into a database table for us. We will also explore a powerful built-in admin interface that provides a visual way of managing all aspects of a Django project, such as users and making changes to model data.

In this chapter, we will cover the following topics:

- Creating our first model
- Installing Pillow
- Managing migrations
- Accessing the Django admin interface
- Configuring image upload
- Serving stored images
- Adding a movie model to admin

Technical requirements

In this chapter, we will be using Python 3.10+. Additionally, we will be using the **VS Code** editor in this book, which you can download from `https://code.visualstudio.com/`.

The code for this chapter is located at `https://github.com/PacktPublishing/Django-5-for-the-Impatient-Second-Edition/tree/main/Chapter05/moviesstore`.

The CiA video for this chapter can be found at `https://packt.link/HEeUM`

Creating our first model

A **Django model** is a Python class that represents a database table. Models are used to define the structure and behavior of the data that will be stored in the database. Each model class typically corresponds to a single database table, and each instance of the class represents a specific row in that table. More information about Django models can be found here: `https://docs.djangoproject.com/en/5.0/topics/db/models/`.

We can create models such as Movie, Review, and Order, and Django turns these models into a database table for us.

Here are the Django model basics:

- Each model is a class that extends `django.db.models.Model`

- Each model attribute represents a database column

- With all of this, Django provides us with a set of useful methods to **create, update, read, and delete** (**CRUD**) model information from a database

Creating a Movie model

Our first model will be a Movie. We can create models in each of the project apps. Movie seems to be more related to the `movies` app, so we will create the `Movie` model there. In `/movies`, we have the `models.py` file, where we create our models for the `movies` app. Open that file and place the following lines of code:

```
from django.db import models

class Movie(models.Model):
    id = models.AutoField(primary_key=True)
    name = models.CharField(max_length=255)
    price = models.IntegerField()
    description = models.TextField()
    image = models.ImageField(upload_to='movie_images/')

    def __str__(self):
        return str(self.id) + ' - ' + self.name
```

Let's explain the previous code:

- First, we import the `models` module, which provides various classes and utilities for defining database models.

- Next, we define a Python class named `Movie`, which inherits from `models.Model`. This means that `Movie` is a Django model class.

- Inside the `Movie` class, we define several fields:

  - `id`: This is an `AutoField` value that automatically increments its value for each new record that's added to the database. The `primary_key=True` parameter specifies that this field is the primary key for the table, uniquely identifying each record.

  - `name`: This is a `CharField` value that represents a string field with a maximum length of 255 characters. It stores the name of the movie.

  - `price`: This is an `IntegerField` value that stores integer values. It represents the price of the movie.

  - `description`: This is a `TextField` value that represents a text field with no specified maximum length. It stores a textual description of the movie.

  - `image`: This is an `ImageField` value that stores image files. The `upload_to` parameter specifies the directory where uploaded images will be stored. In this case, uploaded images will be stored in the `movie_images/` directory within the media directory of the Django project. The media directory is used to store user-uploaded files, such as images, documents, or other media files. This directory is specified in your Django project's settings (we will configure it later in this chapter).

- `__str__`: This is a special method in Python classes that returns a string representation of an object. It concatenates the movie's `id` value (converted into a string) with a hyphen and the movie's name. This method will be useful when we display movies in the Django admin panel later.

> **Note**
>
> Django provides many other model fields to support common types, such as dates, integers, and emails. To have complete documentation of the kinds of types and how to use them, refer to the `Model` field reference in the Django documentation (`https://docs.djangoproject.com/en/5.0/ref/models/fields/`).

Installing Pillow

Because we're using images, we need to install Pillow (`https://pypi.org/project/pillow/`), which adds image-processing capabilities to our Python interpreter.

In the Terminal, stop the server and do the following:

- For macOS, run the following command:

  ```
  pip3 install pillow
  ```

- For Windows, run the following command:

```
pip install pillow
```

Now that Pillow has been installed, let's learn how to manage Django migrations.

Managing migrations

Django migrations is a feature of Django that allows you to manage changes to your database schema – that is, changes to the structure of your database tables and the data within them – over time, as your Django project evolves.

When you define models in Django, you're essentially defining the structure of your database tables. However, as your project grows and changes, you might need to make alterations to these models, such as adding new fields, removing fields, or modifying existing fields. Django migrations provide a way to propagate these changes to your database schema in a controlled and consistent manner (as a version control system).

To work with migrations, we must apply the default migrations, create custom migrations, and apply custom migrations.

Applying the default migrations

Currently, note a message in the Terminal when you run the server:

```
You have 18 unapplied migration(s). Your project may not work properly
until you apply the migrations for app(s): admin, auth, contenttypes,
sessions.
Run 'python manage.py migrate' to apply them.
```

As per the message instructions, stop the server and do the following (remember to be located in the `moviesstore` folder that contains the `manage.py` file):

- For macOS, run the following command:

```
python3 manage.py migrate
```

- For Windows, run the following command:

```
python manage.py migrate
```

The `migrate` command creates an initial database based on Django's default settings. Note that there is a `db.sqlite3` file in the project root folder. This file represents our SQLite database. It's created the first time we run `migrate` or `runserver`.

In the previous case, the `migrate` command applied 18 default migrations (as shown in *Figure 5.1*). Those migrations were defined by some default Django apps – `admin`, `auth`, `contenttypes`, and `sessions`. These apps are loaded in the `INSTALLED_APPS` variable in the `moviesstore/settings.py` file.

So, the `migrate` command runs the migrations of all the installed apps. Note that `INSTALLED_APPS` also loads the `movies` app. However, no migrations were applied for the `movies` app. This is because we haven't generated the migrations for the `movies` app:

```
Operations to perform:
  Apply all migrations: admin, auth, contenttypes, sessions
Running migrations:
  Applying contenttypes.0001_initial... OK
  Applying auth.0001_initial... OK
  Applying admin.0001_initial... OK
  Applying admin.0002_logentry_remove_auto_add... OK
  Applying admin.0003_logentry_add_action_flag_choices... OK
  Applying contenttypes.0002_remove_content_type_name... OK
  Applying auth.0002_alter_permission_name_max_length... OK
  Applying auth.0003_alter_user_email_max_length... OK
  Applying auth.0004_alter_user_username_opts... OK
  Applying auth.0005_alter_user_last_login_null... OK
  Applying auth.0006_require_contenttypes_0002... OK
  Applying auth.0007_alter_validators_add_error_messages... OK
  Applying auth.0008_alter_user_username_max_length... OK
  Applying auth.0009_alter_user_last_name_max_length... OK
  Applying auth.0010_alter_group_name_max_length... OK
  Applying auth.0011_update_proxy_permissions... OK
  Applying auth.0012_alter_user_first_name_max_length... OK
  Applying sessions.0001_initial... OK
```

Figure 5.1 – Applying default Django migrations

Creating custom migrations

Currently, we've defined a `Movie` model inside the `movies` app. Based on that model, we can create our own migrations. To create the migrations for the `movies` app, we need to run the `makemigrations` command in the terminal:

- For macOS, run the following command:

```
python3 manage.py makemigrations
```

- For Windows, run the following command:

```
python manage.py makemigrations
```

The previous command creates migration files based on the models that we've defined in our Django apps (see *Figure 5.2*):

```
Migrations for 'movies':
  movies\migrations\0001_initial.py
    - Create model Movie
```

Figure 5.2 – Executing the makemigrations command

The migrations are stored in the corresponding app-level migrations folder. For now, we have only defined the Movie model inside the movies app. So, this command generates the migration file for the Movie model inside the movies/migrations/ folder (see *Figure 5.3*):

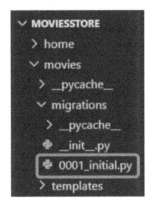

Figure 5.3 – Generated migrations for the movies app

If we change the Movie model or create new models, we need to execute the makemigrations command again. This command will create new migration files that will serve as a version control of our database schema.

Note that the migration file was created, but the database hasn't been updated yet.

Applying custom migrations

After running makemigrations, you typically need to run migrate to apply those migrations to the database and make the corresponding changes. Now, execute the following in the Terminal:

- For macOS, run the following command:

```
python3 manage.py migrate
```

- For Windows, run the following command:

```
python manage.py migrate
```

As shown in *Figure 5.4*, we applied the `movies` app migrations:

```
Operations to perform:
  Apply all migrations: admin, auth, contenttypes, movies, sessions
Running migrations:
  Applying movies.0001 initial... OK
```

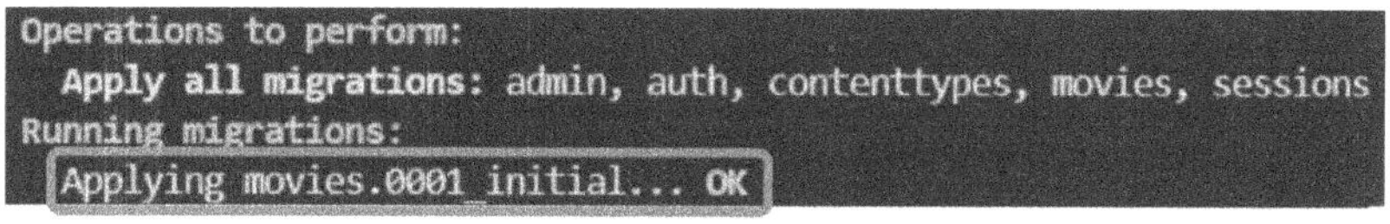

Figure 5.4 – Applying the movies app migrations

In summary, each time you make changes to a model file, you have to do the following:

- For macOS, run the following command:

```
python3 manage.py makemigrations
python3 manage.py migrate
```

- For Windows, run the following command:

```
python manage.py makemigrations
python manage.py migrate
```

But how do we access our database and view what's inside? For that, we use a powerful tool in Django called the admin interface. We'll discuss this in the next section.

Accessing the Django admin interface

To access our database, we have to go into the Django admin interface. Remember that there is an `admin` path in `/moviesstore/urls.py`

```
...
urlpatterns = [
    path('admin/', admin.site.urls),
    path('', include('home.urls')),
    path('movies/', include('movies.urls')),
]
```

If you go to `localhost:8000/admin`, you'll be taken to the admin site, as shown in *Figure 5.5*:

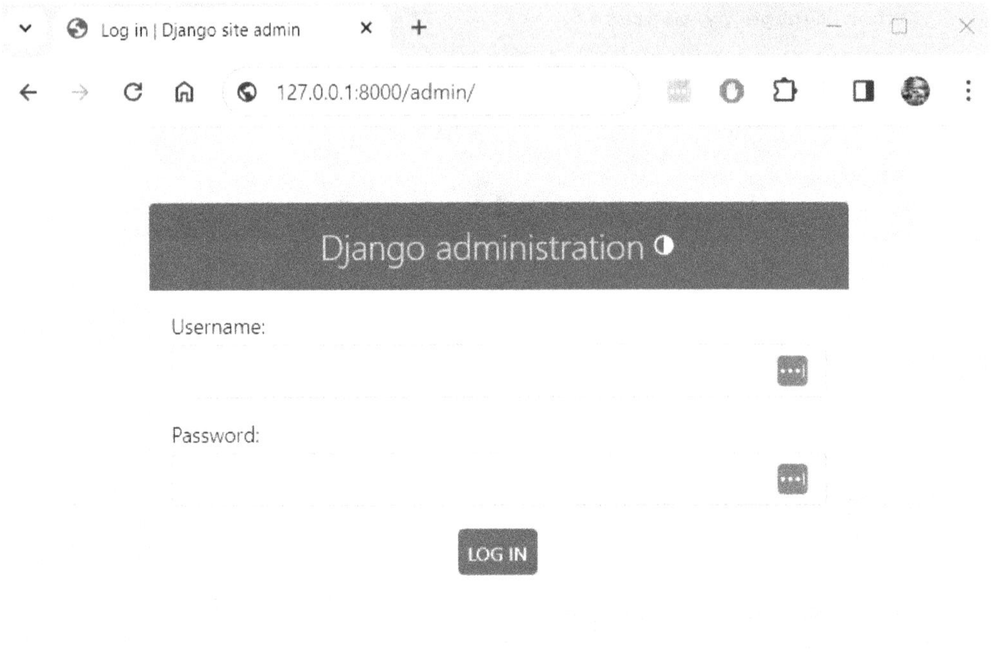

Figure 5.5 – Admin page

Django has a powerful built-in admin interface that provides a visual way of managing all aspects of a Django project – for example, users, movies, and more.

With what username and password do we log in to the admin interface? For this, we have to create a superuser in the Terminal.

Creating a superuser

Let's create a superuser to access the admin panel. In the Terminal, stop the server and do the following:

- For macOS, run the following command:

```
python3 manage.py createsuperuser
```

- For Windows, run the following command:

```
python manage.py createsuperuser
```

You will then be asked to specify a username, email, and password. Note that anyone can access the admin path on your site, so make sure that your password is something secure. After creating the superuser, you should get a message like this from the Terminal:

```
Superuser created successfully.
```

Restoring your superuser password

If you wish to change your password later, you can run the following commands:

- Here's the command for macOS:

  ```
  python3 manage.py changepassword <username>
  ```

- Here's the command for Windows:

  ```
  python manage.py changepassword <username>
  ```

Accessing the admin panel

Now, start the server again and log in to admin with the username you just created, as shown in *Figure 5.6*:

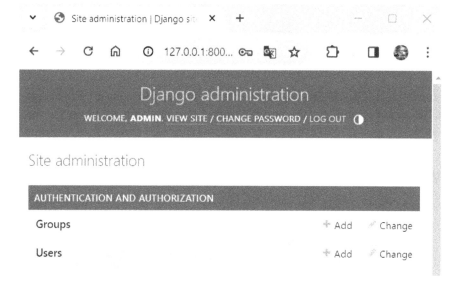

Figure 5.6 – Site administration page

Under **Users**, you'll see the user you've just created, as shown in *Figure 5.7*:

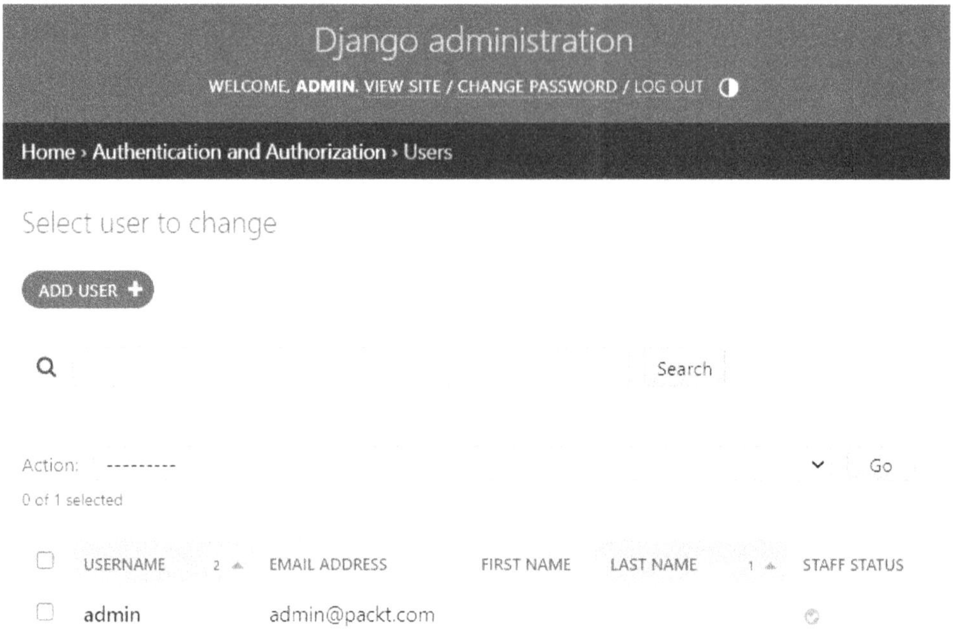

Figure 5.7 – The Users admin page

You can add additional user accounts here for your team.

Currently, our Movie model doesn't show up in admin. We need to explicitly tell Django what to display in it. Before adding our Movie model in admin, let's configure our project so that images can be uploaded.

Configuring image upload

We have to configure where we wish to store our images when we add them. First, in /moviesstore/ settings.py, add the following in bold at the end of the file:

```
...

MEDIA_ROOT = os.path.join(BASE_DIR, 'media')
MEDIA_URL = '/media/'
```

Let's explain the previous code:

- MEDIA_ROOT: This variable specifies the filesystem path to the directory where uploaded media files will be stored. Here, BASE_DIR is a variable that represents the base directory of the Django project, and 'media' is the subdirectory within BASE_DIR where media files will be stored. So, MEDIA_ROOT will be set to a path like /your_project_folder/media.

- MEDIA_URL: This variable specifies the URL prefix that will be used to serve media files from the web server. In this code, it's set to '/media/', meaning that media files uploaded to the Django application will be accessible via URLs starting with /media/. For example, if you upload an image named example.jpg, it might be accessible at a URL like http://localhost:8000/media/example.jpg.

With that, the server has been configured for image upload. So, let's learn how to serve those images.

Serving the stored images

Next, to enable the server to serve the stored images, we have to modify the /moviesstore/urls.py file and add the following in bold:

```
...
from django.conf.urls.static import static
from django.conf import settings

urlpatterns = [
    path('admin/', admin.site.urls),
    path('', include('home.urls')),
    path('movies/', include('movies.urls')),
]

urlpatterns += static(settings.MEDIA_URL,
    document_root=settings.MEDIA_ROOT)
```

With this, you can serve the media files stored in the MEDIA_ROOT directory when the MEDIA_URL URL prefix is accessed.

> **Note**
> It's important to stop the server and run the server again to apply the previous changes.

Now that the image configuration is done, let's add movies to the admin panel.

Adding a movie model to the admin panel

We are now ready to create movies from the admin panel and store the images in our Django project. We will add the `Movie` model to the admin panel, and we will create movies.

Adding the Movie model to the admin panel

To add the `Movie` model to the admin panel, go back to `/movies/admin.py` and register our model by adding the following in **bold**:

```
from django.contrib import admin
from .models import Movie


admin.site.register(Movie)
```

When you save your file, stop the server, run the server, and go back to `/admin`. The `Movie` model will now appear (as shown in *Figure 5.8*):

Figure 5.8 – Admin page with movies available

Try adding a `movie` object by clicking +**Add**. You will be brought to the **Add movie** form, as shown in *Figure 5.9*:

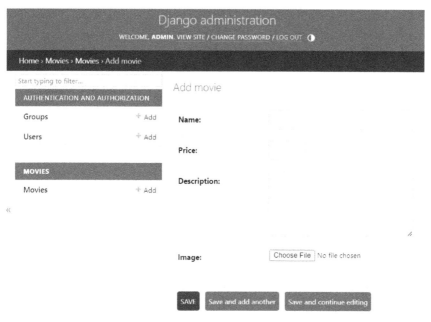

Figure 5.9 – The Add movie form

Try adding a movie and hit **Save**. Your movie object will be saved to the database and reflected in the admin page, as shown in *Figure 5.10*:

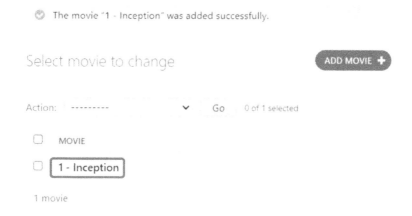

Figure 5.10 – Movies admin page

Note that the admin panel shows the movie's information as a combination of the movie's ID with a hyphen and the movie's name. That's because we defined the `Movie` model's `__str__` method to work like that.

You can also see the movie image in `/moviesstore/media/movie_images/<image file>.jpg`. *Figure 5.11* shows an image called `inception.jpg` stored in the previous folder:

Figure 5.11 – Location of stored movie images

Whenever you upload a movie image, it will be stored in the previous folder. With that, we've configured our project so that it can store and serve images.

Summary

Models are essential for working with databases in Django. In this chapter, we learned about the fundamentals of Django models and created a `Movie` model. We also learned how to use the Django admin interface and how to create movies. In the next chapter, we'll learn how to extract and display the movies stored in our database on our site.

6

Collecting and Displaying Data from the Database

In the previous chapters, movie information was gathered using dummy data implemented within a Python list. While this approach served as a good initial attempt to display movie information, it doesn't scale well. If we want to add a new movie or edit an existing one, we will need to modify our Python code. This chapter focuses on the process of refactoring both the movies and individual movie pages to retrieve and present information directly from the database. With this approach, if we need to add new movies or modify existing ones, we can simply access the admin panel without the need to modify the Python code. Additionally, we will implement a new movie search functionality.

In this chapter, we will cover the following topics:

- Removing the movies' dummy data
- Updating the movie listings page
- Updating the listing of an individual movie page
- Implementing a search movie functionality

By the end of this chapter, you will know how to collect and display information from the database.

Technical requirements

In this chapter, we will use Python 3.10+. Additionally, we will use the **VS Code** editor in this book, which you can download from `https://code.visualstudio.com/`.

The code for this chapter is located at `https://github.com/PacktPublishing/Django-5-for-the-Impatient-Second-Edition/tree/main/Chapter06/moviesstore`.

The CiA video for this chapter can be found at `https://packt.link/mZUvA`

Removing the movies' dummy data

The first step to extract database data is to remove the movies' dummy data. In /movies/views. py, remove the movies variable, as shown in the following in **bold**:

```
from django.shortcuts import render

movies = {
    {
        'id': 1, 'name': 'Inception', 'price': 12,
        'description': 'A mind-bending heist thriller.'
    },
    {
        'id': 2, 'name': 'Avatar', 'price': 13,
        'description': 'A journey to a distant world and the battle
for resources.'
    },
    {
        'id': 3, 'name': 'The Dark Knight', 'price': 14,
        'description': 'Gothams vigilante faces the Joker.'
    },
    {
        'id': 4, 'name': 'Titanic', 'price': 11,
        'description': 'A love story set against the backdrop of the
sinking Titanic.',
    },
}
...
```

We don't need this variable anymore, as we will extract the movie information from the database. Also, remember to access the admin panel and create a few movie objects.

Now that we have removed the dummy data, let's update the way we list movies.

Updating the movie listings page

Now, let's update the code to extract movie information from the database. We will need to, first, update the index function; second, update the movies.index template; and third, add a custom CSS class.

Updating index function

In /movies/views.py, add the following in bold:

```
from django.shortcuts import render
from .models import Movie
```

```
def index(request):
    template_data = {}
    template_data['title'] = 'Movies'
    template_data['movies'] = Movie.objects.all()
    return render(request, 'movies/index.html',
                  {'template_data': template_data})

...
```

Let's explain the previous code:

- We import the `Movie` model from the `models` file. We will use this model to access database information.

- We collect all movies from the database by using the `Movie.objects.all()` method. `Movie.objects` is a manager in Django that serves as the default interface to query the database table associated with the model. It provides various methods to perform database operations such as creating, updating, deleting, and retrieving objects. The `all()` method fetches all objects from the database table represented by the model. Remember that we previously collected the movie information by using the `movies` variable; now, we use the `Movie` Django model.

> **Note**
>
> Django offers several methods to manipulate and access database information. You can find more of these methods here: `https://docs.djangoproject.com/en/5.0/topics/db/queries/`.

Updating the movies.index template

In `/movies/templates/movies/index.html`, add the following in bold:

```
...
{% for movie in template_data.movies %}
<div class="col-md-4 col-lg-3 mb-2">
  <div class="p-2 card align-items-center pt-4">
    <img src="{{ movie.image.url }}"
      class="card-img-top rounded img-card-200">
    <div class="card-body text-center">
      <a href="{% url 'movies.show' id=movie.id %}"
        class="btn bg-dark text-white">
        {{ movie.name }}
      </a>
```

```
        </div>
      </div>
    </div>
    {% endfor %}
    ...
```

We have removed the default image, and now we will show the specific image for each movie. Therefore, we will include a custom CSS class to display the images with the same proportion. We will add this CSS class next.

Adding a custom CSS class

In `/moviesstore/static/css/style.css`, add the following in bold at the end of the file:

```
  ...

.img-card-200 {
  width: fit-content;
  max-height: 200px;
}
```

Now, save those files, run the server, and go to `http://localhost:8000/movies`; you should see the movies page, which extracts information from the database (*Figure 6.1*).

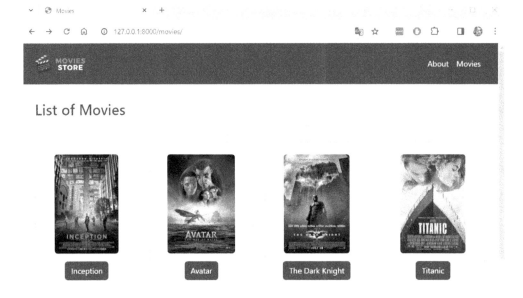

Figure 6.1 – The movies page

The movies page now lists movies from the database; let's complete this process by modifying the individual movie pages.

Updating the listing of an individual movie page

Now, let's update the code to extract individual movie information from the database. We will need to, first, update the show function; second, update the `movies.show` template; and third, add a custom CSS class.

Updating show function

In `/movies/views.py`, add the following in bold:

```
...

def show(request, id):
    movie = Movie.objects.get(id=id)

    template_data = {}
    template_data['title'] = movie.name
    template_data['movie'] = movie
    return render(request, 'movies/show.html',
                  {'template_data': template_data})
```

Let's explain the previous code:

- We use the `Movie.objects.get(id=id)` method to retrieve a specific movie based on its `id`. Remember that `id` is passed by the URL and received as a parameter in the `show` function.

- We now access `movie.name` as an object attribute. Previously, we accessed the name as a key (`movie['name']`), since the dummy data variable stored dictionaries.

Updating the movies.show template

In `/movies/templates/movies/show.html`, add the following in bold:

```
    ...
    <div class="col-md-6 mx-auto mb-3">
      <h2>{{ template_data.movie.name }}</h2>
      <hr />
      <p><b>Description:</b> {{
        template_data.movie.description }}</p>
      <p><b>Price:</b> ${{
        template_data.movie.price }}</p>
```

```
    </div>
    <div class="col-md-6 mx-auto mb-3 text-center">
      <img src="{{ template_data.movie.image.url }}"
        class="rounded img-card-400" />
    </div>
    ...
```

Similar to the previous code, we now show the specific movie image and use a custom CSS class to display movie images with the same proportion.

Adding a custom CSS class

In /moviesstore/static/css/style.css, add the following in bold at the end of the file:

```
    ...


.img-card-400 {
  width: fit-content;
  max-height: 400px;
}
```

Now, save those files, run the server, and go to a specific movie at http://localhost:8000/ movies/1; you should see the individual movie page, which extracts movie information from the database (*Figure 6.2*).

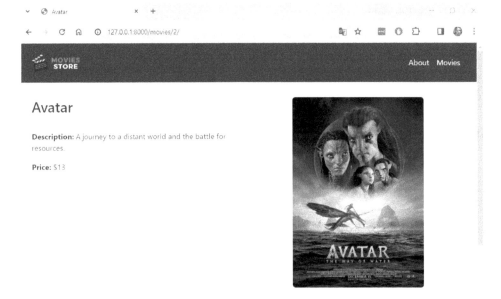

Figure 6.2 – An individual movie page

We are now listing movies and individual movies from the database. Finally, let's include a new functionality to be able to search movies.

Implementing a search movie functionality

Let's finalize this chapter by implementing a search movie functionality. We will need to, first, update the movies.index template, and second, update the index function.

Updating the movies.index template

In /movies/templates/movies/index.html, add the following in bold:

```
...
<div class="col mx-auto mb-3">
  <h2>List of Movies</h2>
  <hr />
  <p class="card-text">
    <form method="GET">
      <div class="row">
        <div class="col-auto">
          <div class="input-group col-auto">
            <div class="input-group-text">
              Search</div>
            <input type="text" class="form-control"
              name="search">
          </div>
        </div>
        <div class="col-auto">
          <button class="btn bg-dark text-white"
            type="submit">Search</button>
        </div>
      </div>
    </form>
  </p>
</div>
</div>
...
```

We have created an HTML form that allows users to perform a search operation. This form will direct to the current URL route and send the search information by the URL. For example, if we search for Avatar, it will direct us to http://localhost:8000/movies/?search=Avatar.

Updating index function

In /movies/views.py, add the following in bold:

...

```
def index(request):
    search_term = request.GET.get('search')
    if search_term:
        movies =
        Movie.objects.filter(name__icontains=search_term)
    else:
        movies = Movie.objects.all()

    template_data = {}
    template_data['title'] = 'Movies'
    template_data['movies'] = movies
    return render(request, 'movies/index.html',
                  {'template_data': template_data})
```

The index function has changed. Now, it will retrieve all movies if the search parameter is not sent in the current request, or it will retrieve specific movies based on the search parameter. Let's explain the previous code.

- We retrieve the value of the search parameter by using the request.GET.get('search') method and assign that value to the search_term variable. Here, we capture the search input value submitted through the form defined in the previous section.

- If search_term is not empty, we filter movies where the name contains search_term. The __icontains lookup is used for a case-insensitive containment search.

- If search_term is empty, we retrieve all movies from the database without applying any filters.

- Finally, we pass the extracted movies to the template_data dictionary.

Now, save those files, run the server, go to http://localhost:8000/movies, enter a search term, and submit the form; you should see the movies that match the search term (*Figure 6.3*).

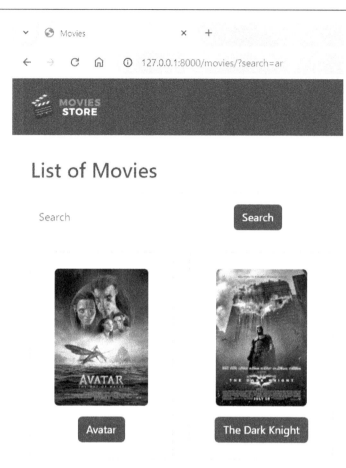

Figure 6.3 – The movies page with a custom search

We have refactored our *Movies Store* code to work with the database instead of dummy data. This strategy enables us to include new movies or edit existing ones without modifying our Python code. Additionally, the addition of the search functionality has helped us understand how to filter different data in Django and enhanced the project's features.

Summary

In this chapter, we learned how to extract information from the database. We learned different Django model methods, such as `all`, `get`, and `filter`, and how they can be used to retrieve different kinds of information. We refactored the movies and individual movie pages to collect information from the database and learned how to implement search functionality.

In the next chapter, we will go deeper into understanding how the database works.

7

Understanding the Database

The previous chapters showed us how to use Django models to persist and retrieve data from a database. In this chapter, we will explore how databases work in Django. We will utilize a database viewer to examine how Django manages various information and stores it. Additionally, we will learn how to customize the Django admin panel and switch between database engines.

In this chapter, we will cover the following topics:

- Understanding the database viewer
- Customizing the Django admin panel
- Switching to a MySQL database

By the end of this chapter, you will understand how the database works, how to visualize database information, and how to switch to a different database engine.

Technical requirements

In this chapter, we will use Python 3.10+. Additionally, we will use the **VS Code** editor in this book, which you can download from `https://code.visualstudio.com/`.

The code for this chapter is located at `https://github.com/PacktPublishing/Django-5-for-the-Impatient-Second-Edition/tree/main/Chapter07/moviesstore`.

The CiA video for this chapter can be found at `https://packt.link/wD2bK`

Understanding the database viewer

Let's take some time to understand how the database works. The objects are stored in the `db.sqlite3` file. If you click on it, it is not very readable. However, you can view such SQLite files with a SQLite Viewer; just google `SQLite Viewer` for a list of them. One example is `https://inloop.github.io/sqlite-viewer/`.

Drag and drop your db.sqlite3 file into the previous link (over the SQLite Viewer), and you will see the different tables in the database (as shown in *Figure 7.1*):

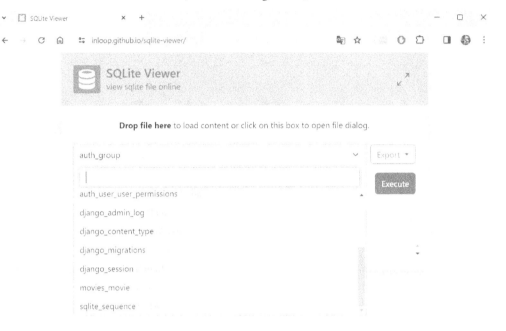

Figure 7.1 – Opening db.sqlite3 in SQLite Viewer

You can see the table of the model we have created – that is, movie. Note that the actual name of the table is determined by combining the name of the app with the name of the model. For example, if your app is named movies and your model is named Movie, the corresponding table name would be movies_movie. This naming convention helps Django differentiate between tables belonging to different apps and models within those apps.

There are also other tables, such as django_session, because of the different apps that are installed for functions such as sessions and authentications.

Select a table (for example, `movies_movie`), and you should be able to see its rows (*Figure 7.2*).

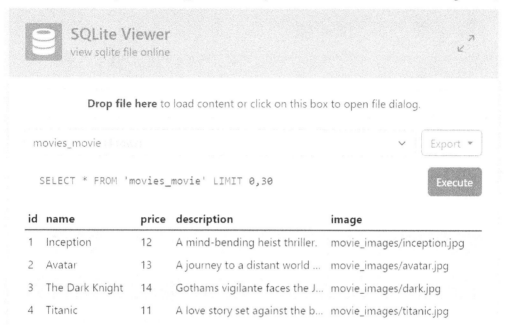

Figure 7.2 – Selecting a table in SQLite Viewer

Hopefully, this lets you appreciate what goes on behind the scenes in a Django database. Currently, we are using an SQLite database. However, what if we want to switch to some other database engines? Django officially supports the following databases – PostgreSQL, MariaDB, MySQL, Oracle, and SQLite.

> **Note**
>
> In addition to the officially supported databases, there are backends provided by third parties that allow you to use other databases with Django, such as CockroachDB, Firebird, Google Cloud Spanner, Microsoft SQL Server, Snowflake, TiDB, and YugabyteDB. You can find more information here: `https://docs.djangoproject.com/en/5.0/ref/databases/#third-party-notes`.

To switch to another database engine, go to `/moviereviews/settings.py` and make changes to the lines in **bold**:

```
...
DATABASES = {
    'default': {
        'ENGINE': 'django.db.backends.sqlite3',
        'NAME': BASE_DIR / 'db.sqlite3',
```

```
        }
    }
    ...
```

You can still create your models as normal, and the changes are handled by Django behind the scenes.

In the book, we use SQLite because it is the simplest. Django uses SQLite by default, and it's a great choice for small projects. It runs off a single file and doesn't require complex installation. In contrast, the other options involve some complexity to configure them properly. We will see at the end of this chapter how to configure a more robust database.

Customizing the Django admin panel

The Django admin panel is a powerful built-in feature of Django that automatically generates a user-friendly interface to manage our application's data models. This is a great feature of Django that many other frameworks don't offer.

Figure 7.3 shows the current movies admin page.

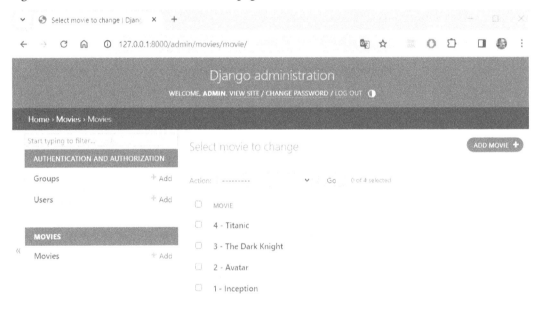

Figure 7.3 – The movies admin page

The admin panel may seem very rigid, but fortunately, Django allows us to customize it according to our needs. Let's apply two customizations to the movies admin page – first, ordering movies by name, and second, allowing searches by name.

Ordering movies by name

In /movies/admin.py, add the following in bold:

```
from django.contrib import admin
from .models import Movie

class MovieAdmin(admin.ModelAdmin):
    ordering = ['name']

admin.site.register(Movie, MovieAdmin)
```

Let's explain the previous code:

- We created a MovieAdmin class that inherits from admin.ModelAdmin. This defines a custom admin class that allows you to customize the behavior of the admin interface for the Movie model.

- We set an ordering attribute. This attribute sets the default ordering of the movie objects in the admin interface. In our case, it specifies that the movies should be ordered by their name field.

- Finally, we registered the Movie model with the custom admin class, MovieAdmin. This tells Django to use the MovieAdmin class to customize the admin interface for the Movie model.

Now, save your file, go back to /admin, and navigate to the movies page. You will see the movie objects ordered by name (as shown in *Figure 7.4*):

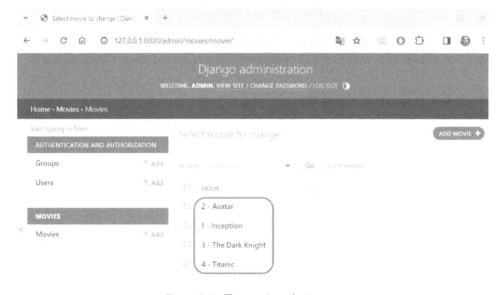

Figure 7.4 – The movies admin page

Allowing searches by name

In /movies/admin.py, add the following in bold:

```
from django.contrib import admin
from .models import Movie

class MovieAdmin(admin.ModelAdmin):
    ordering = ['name']
    search_fields = ['name']

admin.site.register(Movie, MovieAdmin)
```

We added a search_fields attribute that specifies that only the name field of the Movie model is searchable in the admin interface. This means that users can enter keywords into a search box provided by the admin interface, and Django will filter the list of movie objects based on whether the entered keywords match any part of the movie names.

Now, save your file, go back to /admin, and navigate to the movies page. You will see the new search box available (as shown in *Figure 7.5*):

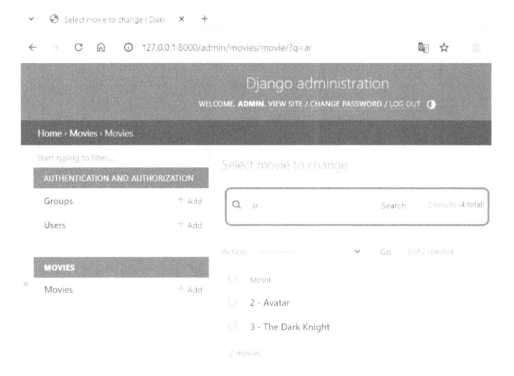

Figure 7.5 – The movies admin page with a search box

> **Note**
>
> As you saw, it is very easy to apply some customizations with very few lines of code. If you want to explore some additional customization, check out this link: `https://docs.djangoproject.com/en/5.0/ref/contrib/admin/`.

Let's finalize this chapter by understanding how to switch to a different database.

Switching to a MySQL database

As we earlier mentioned, we use SQLite throughout this book because it is the simplest. However, we will explain how to switch to a more robust database engine called MySQL.

> **Note**
>
> The book code is based on SQLite, so the changes in this section are optional and won't be reflected either in the GitHub book repository or in upcoming chapters.

MySQL is a popular open source SQL database management system developed by Oracle. There are several different ways to install MySQL. For this section, we will install MySQL and a MySQL administration tool called *phpMyAdmin*. Both tools can be found in a development environment called XAMPP, so let's install that.

XAMPP is a popular PHP development environment. It is a free Apache distribution containing MySQL, PHP, and Perl. As previously mentioned, XAMPP also includes *phpMyAdmin*. If you don't have XAMPP installed, go to `https://www.apachefriends.org/download.html`, download it, and install it.

To switch to a MySQL database, we will need to follow these steps:

1. Configuring the MySQL database.
2. Configuring our project to use the MySQL database.
3. Running the migrations.

Configuring the MySQL database

Execute XAMPP, and then start the Apache module (**1**), start the MySQL module (**2**), and click the MySQL **Admin** button (in the MySQL module) (**3**), which will take us to the *phpMyAdmin* application (as shown in *Figure 7.6*):

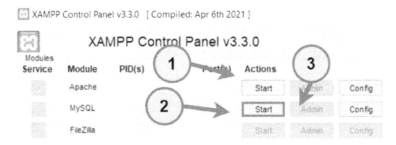

Figure 7.6 – Starting the MySQL module in XAMPP

In the *phpMyAdmin* application, enter your username and password. The default values are root (for the username) and an empty password (*Figure 7.7*):

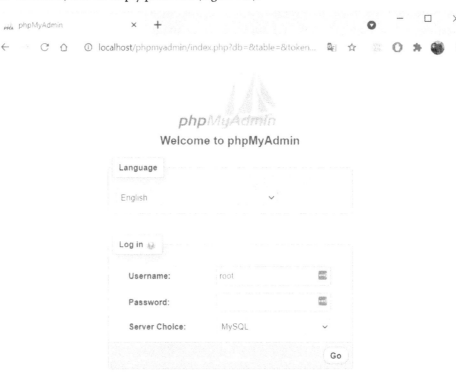

Figure 7.7 – XAMPP phpMyAdmin application

Once you have logged in to *phpMyAdmin*, click the Databases tab (**1**), enter the database name `moviesstore` (**2**), and click the **Create** button (**3**) (as shown in *Figure 7.8*).

Figure 7.8 – Database creation

Configuring our project to use the MySQL database

First, we need to install a package called PyMySQL. PyMySQL is an interface to connect to a MySQL database from Python. Go to the terminal and run the following commands:

- For macOS, run this:

```
pip3 install pymysql
```

- For Windows, run this:

```
pip install pymysql
```

Then, we need to add the following bold lines to the `moviesstore/__init__.py` file:

```
import pymysql
pymysql.install_as_MySQLdb()
```

This `__init__.py` file will be executed when we run the Django project, and the previous two lines import the PyMySQL package into the project.

Finally, we need to modify the database settings to switch to MySQL. In `/moviesstore/settings.py`, modify the `DATABASES` variable to the following in bold:

```
...
DATABASES = {
    'default': {
        'ENGINE': 'django.db.backends.mysql',
        'NAME': 'moviesstore',
        'USER': 'root',
        'PASSWORD': '',
        'HOST': 'localhost',
        'PORT': '3306',
    }
}
...
```

Running the migrations

Since we have switched the database, the new database is empty. So, we need to run the migrations:

- For macOS, run this:

```
python3 manage.py migrate
```

- For Windows, run this:

```
python manage.py migrate
```

Then, we should see the tables in our *phpMyAdmin* application (as shown in *Figure 7.9*).

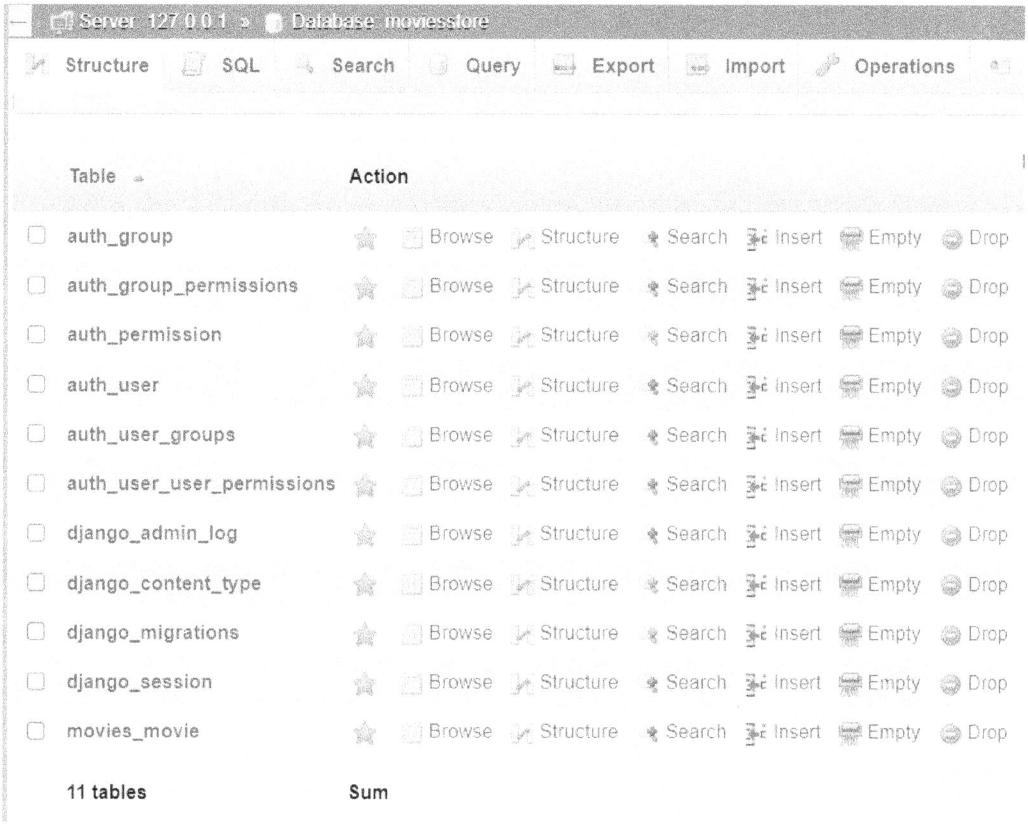

Figure 7.9 – The MySQL database

Finally, we repeat the process of creating a superuser and accessing the admin panel to create some movies.

Summary

We hope that you now better understand how SQLite databases work, how Django supports database management, and how you can customize the Django admin panel. In the next chapter, we will learn how to allow a user to sign up and log in.

8

Implementing User Signup and Login

The next part of our app will concern user authentication, where we allow users to sign up and log in. Implementing user authentication is famously hard. Fortunately, we can use Django's powerful built-in authentication system to take care of the many security pitfalls that could arise if we were to create our own user authentication from scratch.

In this chapter, we will cover the following topics:

- Creating an accounts app
- Creating a basic signup page
- Improving a signup page to handle POST actions
- Customizing UserCreationForm
- Creating a login page
- Implementing the logout functionality

By the end of this chapter, you will know how to implement an authentication system and handle common authentication actions.

Technical requirements

In this chapter, we will use Python 3.10+. Additionally, we will use the VS Code editor in this book, which you can download from https://code.visualstudio.com/.

The code for this chapter is located at https://github.com/PacktPublishing/Django-5-for-the-Impatient-Second-Edition/tree/main/Chapter08/moviesstore.

The CiA video for this chapter can be found at https://packt.link/XmYIk

Creating an accounts app

The complete user authentication system involves a set of functionalities such as a signup, a login, a logout, and some validations. None of these functionalities seem to belong to our *home* app or *movies* app, so let's separate them inside a new app. This new app will be called `accounts`.

We will follow these steps to create and configure the new app:

1. Create an accounts app.
2. Add the accounts app to the settings file.
3. Include an accounts URL file in the project-level URL file.

Let's go through each of these steps in detail in the next few sections.

Creating an accounts app

Navigate to the top `moviesstore` folder (the one that contains the `manage.py` file) and run the following in the terminal:

- For macOS, run the following command:

  ```
  python3 manage.py startapp accounts
  ```

- For Windows, run the following command:

  ```
  python manage.py startapp accounts
  ```

Figure 8.1 shows the new project structure. Verify that it matches your current folder structure.

Figure 8.1 – The MOVIESSTORE project structure containing the accounts app

Now, let's add the accounts app to the settings file.

Adding the accounts app to the settings file

Remember that we must register each newly created app in the settings.py file.

In /moviesstore/settings.py, under INSTALLED_APPS, add the following line in **bold**:

```
...
INSTALLED_APPS = [
    'django.contrib.admin',
    'django.contrib.auth',
    'django.contrib.contenttypes',
    'django.contrib.sessions',
    'django.contrib.messages',
    'django.contrib.staticfiles',
    'home',
    'movies',
    'accounts',
]
...
```

Now, let's include the accounts URL file in our project.

Including the accounts URL file in the project-level URL file

In /moviesstore/urls.py, add the following line that is in bold:

```
...
urlpatterns = [
    path('admin/', admin.site.urls),
    path('', include('home.urls')),
    path('movies/', include('movies.urls')),
    path('accounts/', include('accounts.urls')),
]
...
```

All the URLs defined in the accounts.urls file will contain an accounts/ prefix (as defined in the previous path). We will create the accounts.urls file later.

Now that we have created the accounts app, let's create the first functionality, the signup page.

Creating a basic signup page

The signup page has a complex functionality. We will need to consider many possible scenarios. For now, let's implement a basic signup page. We will refactor and improve this functionality in the upcoming sections.

To implement a basic signup page, we will follow the following steps:

1.   Configure a signup URL.

2.   Define a `signup` function.

3.   Create an accounts signup template.

4.   Add a signup link to the base template.

Let's look at these steps in detail next.

Configuring a signup URL

In `/accounts/`, create a new file called `urls.py`. This file will contain the path relating to the URLs of the accounts app. For now, fill it in with the following code:

```
from django.urls import path
from . import views

urlpatterns = [
    path('signup', views.signup, name='accounts.signup'),
]
```

We defined a `/signup` path, but remember that the project-level URL file defined a `/accounts` prefix for this `urls.py` file. So, if a URL matches the `/accounts/signup` path, it will execute the `signup` function defined in the `views` file. Next, we will implement the `signup` function.

Defining the signup function

In `/accounts/views.py`, add the following lines that are in bold:

```
from django.shortcuts import render
from django.contrib.auth.forms import UserCreationForm

def signup(request):
    template_data = {}
    template_data['title'] = 'Sign Up'
```

```
if request.method == 'GET':
    template_data['form'] = UserCreationForm()
    return render(request, 'accounts/signup.html',
        {'template_data': template_data})
```

Let's explain the code:

- We imported UserCreationForm, which is a built-in form class provided by Django. It is designed to facilitate the creation of user registration forms, specifically to create new user accounts. In Django, we can create our own HTML forms, use some of these Django forms, or even customize the Django forms. We will learn and use all these three approaches in this book.

- We created our template_data variable and assigned it a title.

- Then, we checked whether the current HTTP request method is GET. If it is a GET request, it means that it's a user navigating to the signup form via the localhost:8000/accounts/signup URL, in which case we simply send an instance of UserCreationForm to the template. Finally, we rendered the accounts/signup.html template.

Now, let's continue by creating the signup template.

Creating accounts signup template

In /accounts/, create a templates folder. Then, in /accounts/templates/, create an accounts folder.

Now, in /accounts/templates/accounts/, create a new file, signup.html. For now, fill it in with the following:

```
{% extends 'base.html' %}
{% block content %}
<div class="p-3 mt-4">
  <div class="container">
    <div class="row justify-content-center">
      <div class="col-md-8">
        <div class="card shadow p-3 mb-4 rounded">
          <div class="card-body">
            <h2>Sign Up</h2>
            <hr />
            <form method="POST">
              {% csrf_token %}
              {{ template_data.form.as_p }}
              <button type="submit"
                class="btn bg-dark text-white">Sign Up
              </button>
```

```
            </form>
          </div>
        </div>
      </div>
    </div>
  </div>
</div>
{% endblock content %}
```

Let's explain this code:

- We extend the `base.html` template.

- We define a heading element with the text `Sign Up`.

- We define `form` with its method as `POST`. This means that when the form is submitted, the data will be sent to the current server URL using the HTTP `POST` method.

- Inside the form, we use the DTL `csrf_token` template tag. It generates a **Cross-Site Request Forgery** (**CSRF**) token, which helps prevent CSRF attacks. It ensures that the form submission originates from the same site where the form is rendered. You should use this tag for all your Django forms.

- Inside the form, we render `template_data.form`, which represents the `UserCreationForm` instance passed from the view function. `.as_p` renders the form fields as HTML paragraphs (`<p>`), with each form field wrapped in its own paragraph. By default, `UserCreationForm` contains three form fields – `username`, `password`, and `password confirmation`.

- Inside the form, we include a `submit` button. This button will direct to the current URL using the HTTP `POST` method. Currently, our signup view function only specifies the logic for a `GET` method. Later, we will implement the logic for the `POST` method.

> **Note**
>
> In addition to `form.as_p`, there are other options to render form elements using different HTML tags. You can find more information here: `https://docs.djangoproject.com/en/5.0/ref/forms/api/#output-styles`.

Now, let's finalize by adding the signup link to the base template.

Adding the signup link to the base template

In /moviesstore/templates/base.html, in the header section, add the following lines that are in bold:

```
...
<div class="collapse navbar-collapse"
            id="navbarNavAltMarkup">
  <div class="navbar-nav ms-auto navbar-ml">
    <a class="nav-link"
      href="{% url 'home.about' %}">About</a>
    <a class="nav-link" href=
      "{% url 'movies.index' %}">Movies</a>
    <div class="vr bg-white mx-2 d-none
      d-lg-block"></div>
    <a class="nav-link"
      href="{% url 'accounts.signup' %}">
      Sign Up
    </a>
  </div>
</div>
...
```

Now, save those files, run the server, and go to http://localhost:8000/accounts/signup; you should see the new *signup* page (*Figure 8.2*).

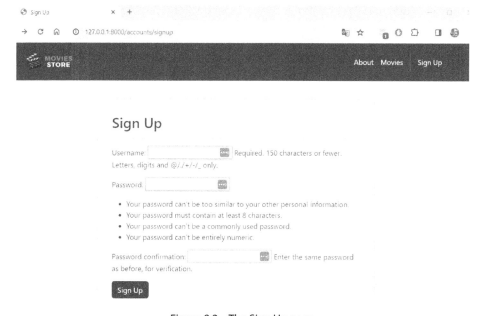

Figure 8.2 – The Sign Up page

Note that if you try to complete and submit the form, it will display an error. This is because we haven't completed the `signup` function.

Improving the signup page to handle POST actions

When a user submits the signup form, we will have to handle the request and create a user in admin. To implement this, we will modify the `signup` function.

In `/accounts/views.py`, add the following lines that are in bold:

```
from django.shortcuts import render
from django.contrib.auth.forms import UserCreationForm
from django.shortcuts import redirect

def signup(request):
    template_data = {}
    template_data['title'] = 'Sign Up'

    if request.method == 'GET':
        template_data['form'] = UserCreationForm()
        return render(request, 'accounts/signup.html',
            {'template_data': template_data})
    elif request.method == 'POST':
        form = UserCreationForm(request.POST)
        if form.is_valid():
            form.save()
            return redirect('home.index')
        else:
            template_data['form'] = form
            return render(request, 'accounts/signup.html',
                {'template_data': template_data})
```

Let's explain this code:

- We import the `redirect` function, which is used to redirect the user to a different URL within the application.

- We add an `elif` section. This section checks whether the HTTP request method is POST, indicating that the form has been submitted.

- Inside the `elif` section, we create an instance of the `UserCreationForm` class, passing the data from the request's POST parameters (`request.POST`) to populate the form fields. This initializes the form with the submitted data.

- The `if form.is_valid()` checks whether the submitted form data is valid, according to the validation rules defined in the `UserCreationForm` class. These validations include that the two password fields match, the password is not common, and the username is unique, among others.

  - If the form data is valid, `form.save()` saves the user data to the database. This means creating a new user account with the provided username and password. Also, we redirect the user to the `home` page based on the URL pattern name.

  - If the form data is not valid, the `else` section is executed, and we pass the form (including the errors) to the template and render the `accounts/signup.html` template again.

Now, run the server, and go to `http://localhost:8000/accounts/signup`. First, try to register a user with two passwords that don't match (*Figure 8.3*).

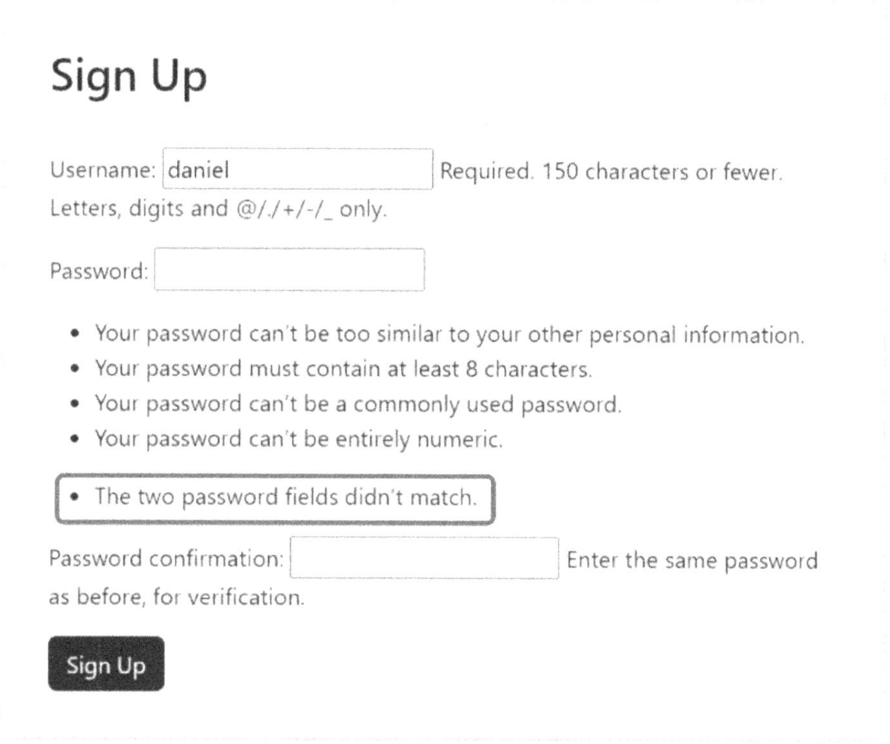

Figure 8.3 – The Sign Up page with errors

Then, try to register a user with proper information, and you should be redirected to the *home* page. Then, go to http://127.0.0.1:8000/admin/, navigate to the *users* section, and you should see the new user registered in the database (*Figure 8.4*).

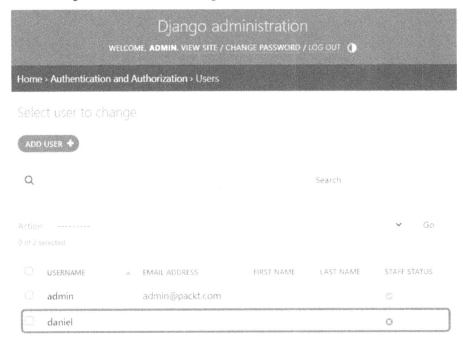

Figure 8.4 – The users admin page

We can now register users. Now, let's customize UserCreationForm.

Customizing UserCreationForm

UserCreationForm currently shows quite a lot of additional help text (included by default) that is cluttering our form. To remedy this, we can customize UserCreationForm (which is a big topic on its own). Here, we will apply some simple modifications to improve the look and feel of our signup page.

To implement these modifications, we will follow these steps:

1. Create CustomUserCreationForm.
2. Update the signup function to use CustomUserCreationForm.
3. Customize the way errors are displayed.

We will undertake each of these steps in detail in the next few subsections.

Creating CustomUserCreationForm

In `/accounts/`, create a new file called `forms.py`. This file will contain the custom forms of the *accounts* app. For now, fill it in with the following code:

```python
from django.contrib.auth.forms import UserCreationForm

class CustomUserCreationForm(UserCreationForm):
    def __init__(self, *args, **kwargs):
        super(CustomUserCreationForm, self).__init__
            (*args, **kwargs)
        for fieldname in ['username', 'password1',
        'password2']:
            self.fields[fieldname].help_text = None
            self.fields[fieldname].widget.attrs.update(
                {'class': 'form-control'}
            )
```

Let's explain the code:

- We import the `UserCreationForm` class from Django's authentication forms.
- We create a new class named `CustomUserCreationForm`, which inherits from `UserCreationForm`, making it a subclass of Django's built-in user creation form.
- We define the class constructor (the `__init__` method). The constructor calls the constructor of the parent class (`UserCreationForm`) through the `super` method.
- Then, we iterate through the fields provided by `UserCreationForm`. These are `'username'`, `'password1'`, and `'password2'`. For each field specified in the loop, we set the `help_text` attribute to `None`, which removes any help text associated with these fields. Finally, for each field specified in the loop, we add the CSS `form-control` class to the field's widget. This is a Bootstrap class that improves the look and feel of the fields.

Next, let's use `CustomUserCreationForm` in our `signup` function.

Updating the signup function to use CustomUserCreationForm

Let's use the new form to improve the look and feel of the signup page.

In `/accounts/views.py`, add the following lines that are in bold:

```python
from django.shortcuts import render
from .forms import CustomUserCreationForm
from django.shortcuts import redirect
```

```
def signup(request):
    template_data = {}
    template_data['title'] = 'Sign Up'

    if request.method == 'GET':
        template_data['form'] = CustomUserCreationForm()
        return render(request, 'accounts/signup.html',
            {'template_data': template_data})
    elif request.method == 'POST':
        form = CustomUserCreationForm(request.POST)
        if form.is_valid():
            form.save()
            return redirect('home.index')
        else:
            template_data['form'] = form
            return render(request, 'accounts/signup.html',
                {'template_data': template_data})
```

In the updated code, we removed the import of UserCreationForm and added the import of CustomUserCreationForm. Then, we replaced the calls of UserCreationForm() with the calls of CustomUserCreationForm().

Now, save those files, run the server, go to http://localhost:8000/accounts/signup, and try to register a user with two passwords that don't match (*Figure 8.5*); you will see that the look and feel have improved.

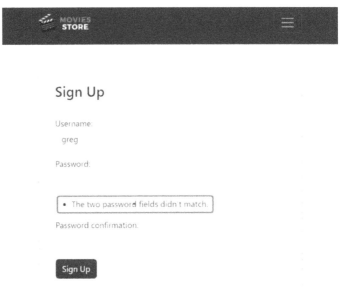

Figure 8.5 – An improved Sign Up page with errors

We can improve the way errors are displayed. So, let's customize this in the next section.

Customizing the way errors are displayed

Let's customize the way Django forms display errors. In /accounts/forms.py, add the following lines that are in bold:

```
from django.contrib.auth.forms import UserCreationForm
from django.forms.utils import ErrorList
from django.utils.safestring import mark_safe

class CustomErrorList(ErrorList):
    def __str__(self):
        if not self:
            return ''
        return mark_safe(''.join([
            f'<div class="alert alert-danger" role="alert">
            {e}</div>' for e in self]))

class CustomUserCreationForm(UserCreationForm):
    ...
```

Let's explain the code:

- We import the `ErrorList` class, which is a default class used to store and display validation error messages associated with form fields.

- We import the `mark_safe` function, which is used to mark a string as safe for HTML rendering, indicating that it doesn't contain any harmful content and should be rendered as-is without escaping.

- We define a new class named `CustomErrorList`, which extends Django's `ErrorList` class. This will be the class to define our custom error look and feel.

- We override the `__str__()` method of the base `ErrorList` class. If the error list is empty (i.e., there are no errors), it returns an empty string, indicating that no HTML should be generated. Otherwise, it defines a custom HTML code that uses `<div>` elements and Bootstrap CSS classes to improve the way the errors are displayed. It also uses the `mark_safe` function to render the code as-is.

Now that we have defined this `CustomErrorList` class, we just need to specify to our forms that we will use it.

In /accounts/views.py, add the following in bold:

```
from django.shortcuts import render
from .forms import CustomUserCreationForm, CustomErrorList
from django.shortcuts import redirect
```

```
def signup(request):
    template_data = {}
    template_data['title'] = 'Sign Up'

    if request.method == 'GET':
        template_data['form'] = CustomUserCreationForm()
        return render(request, 'accounts/signup.html',
            {'template_data': template_data})
    elif request.method == 'POST':
        form = CustomUserCreationForm(request.POST,
            error_class=CustomErrorList)
        ...
```

We imported our CustomErrorList class, and we passed this class as an argument to CustomUserCreationForm. This time, if an error is found when we submit the signup form, the form will use our CustomErrorList class and display the errors with our custom HTML and CSS code.

Now, save those files, run the server, go to http://localhost:8000/accounts/signup, and try to register a user with two passwords that don't match (*Figure 8.6*); you will see that the look and feel have improved.

Figure 8.6 – The Sign Up page with an improved error style

We have improved the look and feel of our errors. Now, let's implement a login page.

Creating a login page

Let's implement the login page. This time, we won't use Django forms; we will create our own HTML form (to learn a new approach). Let's follow the following steps:

1. Configure a login URL.

2. Define the `login` function.

3. Create an accounts login template.

4. Add a link to the base template.

5. Redirect a registered user to the login page.

We'll see these steps to create a login page, in depth, in the next few subsections.

Configuring a login URL

In /accounts/urls.py, add the following path in bold:

```
from django.urls import path
from . import views

urlpatterns = [
    path('signup', views.signup, name='accounts.signup'),
    path('login/', views.login, name='accounts.login'),
]
```

So, if a URL matches the /accounts/login path, it will execute the login function defined in the views file.

Now that we have the new path, let's define the login function.

Defining login function

In /accounts/views.py, add the following lines that are in bold:

```
from django.shortcuts import render
from django.contrib.auth import login as auth_login, authenticate
from .forms import CustomUserCreationForm, CustomErrorList
from django.shortcuts import redirect

def login(request):
    template_data = {}
```

```
        template_data['title'] = 'Login'
        if request.method == 'GET':
            return render(request, 'accounts/login.html',
                {'template_data': template_data})
        elif request.method == 'POST':
            user = authenticate(
                request,
                username = request.POST['username'],
                password = request.POST['password']
            )
            if user is None:
                template_data['error'] =
                    'The username or password is incorrect.'
                return render(request, 'accounts/login.html',
                    {'template_data': template_data})
            else:
                auth_login(request, user)
                return redirect('home.index')

def signup(request):
    ...
```

Let's explain the code:

- We import `login` and `authenticate`. These are used for user authentication. We import `login` with an alias (`auth_login`) to avoid confusion with the `login` function name.

- We create the `login` function. This function defines `template_data` and checks `request.method`.

- For GET requests, the function renders the `accounts/login.html` template.

- For POST requests, the function attempts to authenticate the user using the provided `username` and `password`. If authentication fails, it renders the login template again with an error message. If authentication succeeds, it logs the user in and redirects them to the *home* page.

Now, let's create the template that requires the `login` function.

Creating accounts login template

In `/accounts/templates/accounts/`, create a new file, `login.html`. This file contains the HTML for the login page. For now, fill it in with the following:

```
{% extends 'base.html' %}
{% block content %}
<div class="p-3 mt-4">
```

```html
<div class="container">
  <div class="row justify-content-center">
    <div class="col-md-8">
      <div class="card shadow p-3 mb-4 rounded">
        <div class="card-body">
          <h2>Login</h2>
          <hr />
          {% if template_data.error %}
          <div class="alert alert-danger" role="alert">
            {{ template_data.error }}
          </div>
          {% endif %}
          <form method="POST">
            {% csrf_token %}
            <p>
              <label for="username">Username</label>
              <input id="username" type="text"
                     name="username" required
                     autocomplete="username"
                     class="form-control">
            </p>
            <p>
              <label for="password">Password</label>
              <input id="password" type="password"
                     name="password" required
                     autocomplete="current-password"
                     class="form-control">
            </p>
            <div class="text-center">
              <button type="submit"
                class="btn bg-dark text-white">Login
              </button>
            </div>
          </form>
        </div>
      </div>
    </div>
  </div>
</div>
{% endblock content %}
```

Let's explain the code:

- We extend the `base.html` template and define a heading element with the text `Login`.

- We check whether there is an error and, if so, display it.

- We create an HTML form with a `POST` method and the `csrf_token` token. This form contains two inputs, one for the username and another for the password. It also contains a submit button.

Let's continue by adding the login link to the base template.

Adding the link to the base template

Let's add the login link in the base template. In `/moviesstore/templates/base.html`, in the header section, add the following line that is in bold:

```
...
<div class="navbar-nav ms-auto navbar-ml">
  <a class="nav-link"
    href="{% url 'home.about' %}">About</a>
  <a class="nav-link"
    href="{% url 'movies.index' %}">Movies</a>
  <div class=
    "vr bg-white mx-2 d-none d-lg-block"></div>
  <a class="nav-link"
    href="{% url 'accounts.login' %}">Login</a>
  <a class="nav-link"
    href="{% url 'accounts.signup' %}">Sign Up
  </a>
</div>
...
```

Now, save those files, run the server, and go to `http://localhost:8000/accounts/login`; you will see the new **Login** page (*Figure 8.7*).

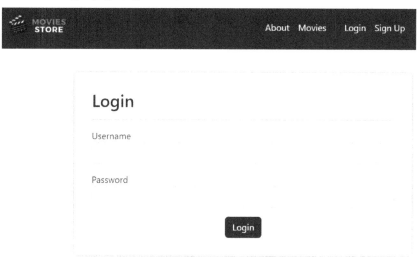

Figure 8.7 – The Login page

Now that we have a login page, let's redirect the user to it when they create an account.

Redirecting a registered user to the login page

Let's finalize this section by redirecting a user who just registered to the **Login** page. In `/accounts/views.py`, add the following in bold:

```
...
def signup(request):
    template_data = {}
    template_data['title'] = 'Sign Up'

    if request.method == 'GET':
        template_data['form'] = CustomUserCreationForm()
        return render(request, 'accounts/signup.html',
            {'template_data': template_data})
    elif request.method == 'POST':
        form = CustomUserCreationForm(request.POST,
            error_class=CustomErrorList)
        if form.is_valid():
            form.save()
            return redirect('accounts.login')
            ...
```

We just modified the redirection to the **Login** page. Try to create a new user, and they should be redirected to the **Login** page.

Let's finalize this chapter by implementing a logout functionality.

Implementing a logout functionality

We'll follow the following steps:

1. Configure a logout URL.
2. Define the `logout` function.
3. Add a link to the base template.

We'll undertake these steps in the upcoming sections.

Configuring a logout URL

In /accounts/urls.py, add the path that is in bold:

```
from django.urls import path
from . import views

urlpatterns = [
    path('signup', views.signup, name='accounts.signup'),
    path('login/', views.login, name='accounts.login'),
    path('logout/', views.logout, name='accounts.logout'),
]
```

Now, if a URL matches the /accounts/logout path, it will execute the logout function defined in the views file.

Defining the logout function

In /accounts/views.py, add the following lines that are in bold:

```
from django.shortcuts import render
from django.contrib.auth import login as auth_login, authenticate,
logout as auth_logout
from .forms import CustomUserCreationForm, CustomErrorList
from django.shortcuts import redirect
from django.contrib.auth.decorators import login_required

@login_required
def logout(request):
    auth_logout(request)
```

```
    return redirect('home.index')

def login(request):
    ...
```

Let's explain the code:

- We import the `logout` function as `auth_logout`. This is used to log a user out.

- We import `login_required`, which is a decorator to ensure that only authenticated users can access specific view functions. A Django **decorator** is a function that wraps another function or method to modify its behavior. Decorators are commonly used for things such as authentication, permissions, and logging.

- We create the `logout` function, which uses the `login_required` decorator. This means that only authenticated users can access this function.

- The `logout` function calls `auth_logout`, which is used to log out the current user. Then, the function redirects the user to the *home* page.

Next, let's add the logout link to the base template.

Adding the link to the base template

In `/moviesstore/templates/base.html`, in the header section, add the following lines that are in bold:

```
...
<a class="nav-link"
  href="{% url 'home.about' %}">About</a>
<a class="nav-link"
  href="{% url 'movies.index' %}">Movies</a>
<div class=
  "vr bg-white mx-2 d-none d-lg-block"></div>
{% if user.is_authenticated %}
<a class="nav-link"
  href="{% url 'accounts.logout' %}">Logout ({{
  user.username }})</a>
{% else %}
<a class="nav-link"
  href="{% url 'accounts.login' %}">Login</a>
<a class="nav-link"
  href="{% url 'accounts.signup' %}">Sign Up
</a>
{% endif %}
...
```

We use a Django template tag that checks whether the user is authenticated (logged in). This validation comes from Django's authentication system. If the user is authenticated, we display the *logout* option (which includes the username). Otherwise, we display the *login* and *sign up* options.

Now, save those files, run the server, and go to `http://localhost:8000/`; you will see how the navbar options change whether the user is logged in or not (*Figure 8.8*).

Figure 8.8 – The home page with the navbar updated

We have completed our user signup, login, and logout system.

Summary

In this chapter, we implemented a complete authentication system. Now, users can sign up, log in, and log out. We also learned how to take advantage of some Django forms, how to create our own HTML forms, and how to handle validations and errors.

In the next chapter, we will implement a movie review system.

9

Letting Users Create, Read, Update, and Delete Movie Reviews

Having implemented the authentication system, it is now time to let logged-in users perform the standard CRUD operations on reviews for movies. This chapter will teach you how to perform complete CRUD operations and how to manage authorizations.

In this chapter, we will cover the following topics:

- Creating a review model
- Creating reviews
- Reading reviews
- Updating a review
- Deleting a review

By the end of the chapter, you will have learned how to create CRUD operations for your models and handle authorizations. You will also recap how to use forms and how to manage different HTTP methods.

Technical requirements

In this chapter, we will be using Python 3.10 or above. Additionally, we will be using the **VS Code** editor in this book, which you can download from `https://code.visualstudio.com/`.

The code for this chapter is located at `https://github.com/PacktPublishing/Django-5-for-the-Impatient-Second-Edition/tree/main/Chapter09/moviesstore`.

The CiA video for this chapter can be found at `https://packt.link/dsqdR`

Creating a review model

To store the movies' review information, we need to create a review Django model and follow the next steps:

1. Create the review model.
2. Apply migrations.
3. Add the review model to the admin panel.

Create the review model

The review information is closely connected to movies. Therefore, we will include this model in the movies app. In the /movies/models.py file, add the following parts that are highlighted in bold:

```python
from django.db import models
from django.contrib.auth.models import User

class Movie(models.Model):
    ...

class Review(models.Model):
    id = models.AutoField(primary_key=True)
    comment = models.CharField(max_length=255)
    date = models.DateTimeField(auto_now_add=True)
    movie = models.ForeignKey(Movie,
        on_delete=models.CASCADE)
    user = models.ForeignKey(User,
        on_delete=models.CASCADE)

    def __str__(self):
        return str(self.id) + ' - ' + self.movie.name
```

Let's explain the preceding code:

- We import the User model from Django's django.contrib.auth.models module.
- We define a Python class named Review, which inherits from models.Model. This means that Review is a Django model class.

- Inside the `Review` class, we define several fields:

  ▪ `id` is an `AutoField`, which automatically increments its value for each new record added to the database. The `primary_key=True` parameter specifies that this field is the primary key for the table, uniquely identifying each record.

  ▪ `comment` is a `CharField`, which represents a string field with a maximum length of 255 characters. It stores the movie review text.

  ▪ `date` is a `DateTimeField`, which is used for date and time data. The `auto_now_add=True` ensures that the date and time are automatically set to the current date and time when the review is created.

  ▪ `movie` is a foreign key relationship to the `Movie` model. A review is associated with a movie. The `on_delete` parameter specifies how to handle the deletion of a movie that a review is associated with. In this case, `on_delete=models.CASCADE` means that if the related movie is deleted, the associated review will also be deleted.

  ▪ `user` is another foreign key relationship but to the `User` model. A review is associated with a user (the person who wrote the review). Similar to the `movie` attribute, `on_delete=models.CASCADE` specifies that if the related user is deleted, the associated review will also be deleted.

- `__str__` is a method that returns a string representation of the review. In this case, it returns a string that is composed of the review ID and the name of the movie associated with the review.

Apply migrations

Now that we have created the `Review` model, let's apply those changes to our database by running the following commands according to your operating system:

- For macOS, run this:

```
python3 manage.py makemigrations
python3 manage.py migrate
```

- For Windows, run this:

```
python manage.py makemigrations
python manage.py migrate
```

Now, you should see something like in *Figure 9.1*:

```
Operations to perform:
  Apply all migrations: admin, auth, contenttypes, movies, sessions
Running migrations:
  Applying movies.0002_review... OK
```

Figure 9.1 – Applying the review migration

Add the review model to the admin panel

To add the `Review` model to `admin`, go back to `/movies/admin.py` and register it by adding the following parts that are highlighted in bold:

```python
from django.contrib import admin
from .models import Movie, Review

class MovieAdmin(admin.ModelAdmin):
    ordering = ['name']
    search_fields = ['name']

admin.site.register(Movie, MovieAdmin)
admin.site.register(Review)
```

When you save your file, stop the server, run the server, and go back to `/admin`. The review model will now show up (as shown in *Figure 9.2*):

Figure 9.2 – The admin page with reviews available

Now that we have created and applied our `Review` model, let's create the functionality to create reviews.

Creating reviews

To allow users to create reviews, we need to follow the next steps:

1. Update the `movies.show` template.
2. Define the `create_review` function.
3. Configure the `create review` URL.

Updating the movies.show template

We will include a form to allow authenticated users to create reviews. This form will be included in the `movies.show` template. In the `/movies/templates/movies/show.html` file, add the following, as presented in bold:

```
...
<p><b>Price:</b> ${{ template_data.movie.price }}</p>
{% if user.is_authenticated %}
<div class="container mt-4">
  <div class="row justify-content-center">
    <div class="col-12">
      <div class="card shadow p-3 mb-4 rounded">
        <div class="card-body">
          <b class="text-start">Create a review
            </b><br /><br />
          <form method="POST" action=
            "{% url 'movies.create_review'
            id=template_data.movie.id %}">
            {% csrf_token %}
            <p>
              <label for="comment">Comment:</label>
              <textarea name="comment" required
              class="form-control"
              id="comment"></textarea>
            </p>

            <div class="text-center">
              <button type="submit"
                class="btn bg-dark text-white">
                Add Review
              </button>
            </div>
```

```
                </form>
              </div>
            </div>
          </div>
        </div>
      </div>
      {% endif %}
    </div>
    <div class="col-md-6 mx-auto mb-3 text-center">
      <img src="{{ template_data.movie.image.url }}"
        class="rounded img-card-400" />
    </div>
    ...
{% endblock content %}
```

Let's explain the preceding code:

- We use the {% if user.is_authenticated %} DTL conditional statement that checks whether the user is authenticated (logged in). If the user is authenticated, the block of HTML code within the if statement will be rendered and displayed.

- We create an HTML form with the POST method and the csrf_token token. This form contains a single input named comment. This input stores the review text. The form also contains a submit button.

- The form is linked to the movies.create_review URL, and it also passes the movie ID to that URL. The movie ID will be used to link the current comment with the movie that it represents.

Defining the create_review function

In /movies/views.py, add the following, as presented in bold:

```
from django.shortcuts import render, redirect
from .models import Movie, Review
from django.contrib.auth.decorators import login_required

def index(request):
    ...

def show(request):
    ...

@login_required
def create_review(request, id):
```

```
if request.method == 'POST' and request.POST['comment']
!= '':
    movie = Movie.objects.get(id=id)
    review = Review()
    review.comment = request.POST['comment']
    review.movie = movie
    review.user = request.user
    review.save()
    return redirect('movies.show', id=id)
else:
    return redirect('movies.show', id=id)
```

Let's explain the preceding code:

- We import the `redirect` function, which is used to redirect the user to a different URL.

- We import the `Review` model, which will be used to create new reviews.

- We import `login_required`, which is used to verify that only logged users can access the `create_review` function. If a guest user attempts to access this function via the corresponding URL, they will be redirected to the login page.

- We create the `create_review` function that handles creating a review.

- The `create_review` takes two arguments: the `request` that contains information about the HTTP request, and the `id`, which represents the ID of the movie for which a review is being created.

- Then, we check whether the request method is POST and the `comment` field in the request's POST data is not empty. If that is TRUE, the following happens:

- We retrieve the movie using `Movie.objects.get(id=id)` based on the provided `id`.

  - We create a new `Review` object.

  - We set the review properties as follows:

    - We set the `comment` based on the comments collected in the form

    - We set the `movie`, based on the retrieved movie from the database

    - We set the `user`, based on the authenticated user who submitted the form.

  - Finally, we save the review to the database and redirect the user to the movie show page.

- In the `else` case, we redirect the user to the movie show page using the `redirect('movies.show', id=id)` code.

Configuring the create review URL

In /movies/urls.py, add the next path as highlighted in bold:

```
from django.urls import path
from . import views

urlpatterns = [
    path('', views.index, name='movies.index'),
    path('<int:id>/', views.show, name='movies.show'),
    path('<int:id>/review/create/', views.create_review,
        name='movies.create_review'),
]
```

Let's analyze the new path. The <int:id> part indicates that this path expects an integer value to be passed from the URL and that the integer value will be associated with a variable named id. The id variable will be used to identify to which movie the review that we want to create is linked. For example, if the form is submitted to movies/1/review/create, it indicates that the new review will be associated with the movie with id=1.

Now save those files, run the server, and go to http://localhost:8000/movies. Click on a specific movie and you will see the form to create reviews (*Figure 9.3*).

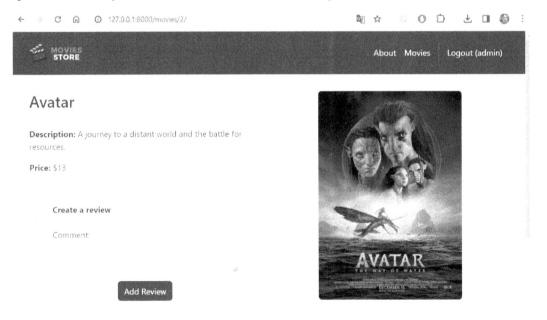

Figure 9.3 – A movie page with the review form

Then, enter a comment and click **Add Review**. A new review should be created, and you should be redirected to the movie show page. Go to the admin panel, click **Reviews**, and you will see the new review there (*Figure 9.4*).

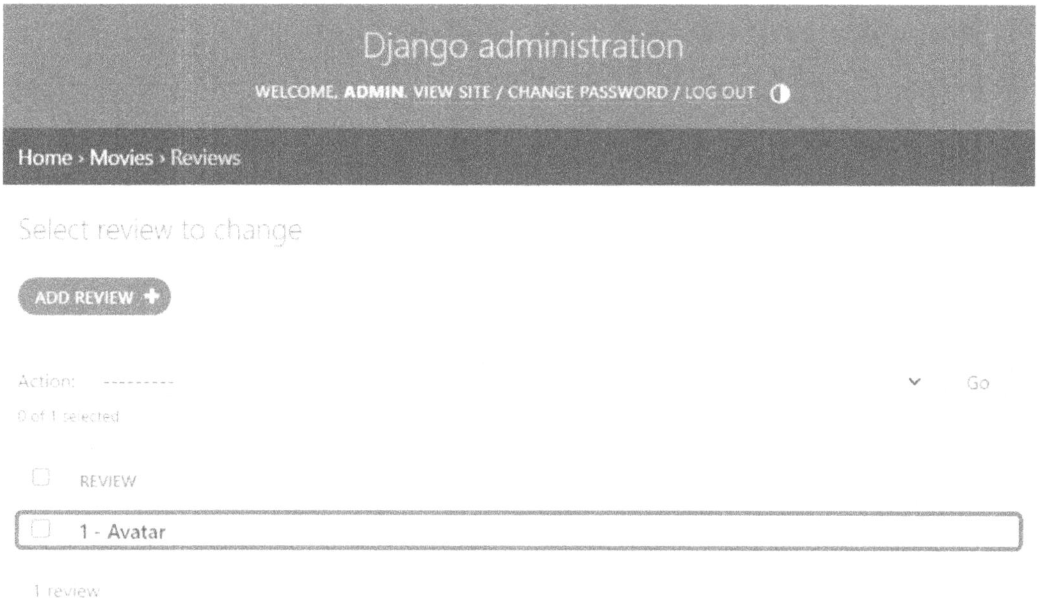

Figure 9.4 – The reviews admin page

Let's now include a functionality to read and list reviews from our web application.

Reading reviews

To be able to read and list reviews, we need to follow the steps that follow:

1. Update the `movies.show` template.
2. Update the `show` function.

Updating the movies.show template

We will list the reviews in the `movies.show` template. In the `/movies/templates/movies/show.html` file, add the following, as highlighted in bold:

```
...
<p><b>Price:</b> ${{ template_data.movie.price }}
  </p>
```

```
<h2>Reviews</h2>
<hr />
<ul class="list-group">
  {% for review in template_data.reviews %}
  <li class="list-group-item pb-3 pt-3">
    <h5 class="card-title">
      Review by {{ review.user.username }}
    </h5>
    <h6 class="card-subtitle mb-2 text-muted">
      {{ review.date }}
    </h6>
    <p class="card-text">{{ review.comment }}</p>
  </li>
  {% endfor %}
</ul>

{% if user.is_authenticated %}
  ...
```

We have added a new section inside the template. This section iterates through the `reviews` and displays the review `date` and `comment`, as well as the username of the user who created the review.

Updating the show function

In `/movies/views.py`, add the following, as highlighted in bold:

```
...

def show(request, id):
    movie =  Movie.objects.get(id=id)
    reviews = Review.objects.filter(movie=movie)

    template_data = {}
    template_data['title'] = movie.name
    template_data['movie'] = movie
    template_data['reviews'] = reviews
    return render(request, 'movies/show.html',
        {'template_data': template_data})

...
```

Let's explain the preceding code.

- We retrieve all review objects that are associated with the movie that we are showing. To do this, we use the `filter` method to limit the query to reviews related to the specific movie.

- We add those reviews to the `template_data` dictionary, which is passed to the `movies/show.html` template.

Now, save those files, run the server, and go to `http://localhost:8000/movies`. Click on a specific movie that contains reviews and you will see the movie information, including its corresponding reviews (*Figure 9.5*).

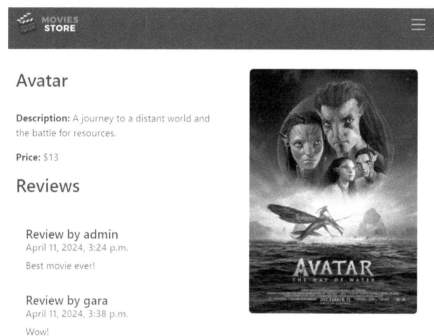

Figure 9.5 – A movie page with reviews

Now, let's move on to updating reviews.

Updating a review

To be able to update reviews, we need to follow these steps:

1. Update the `movies.show` template.
2. Create the `movies edit_review` template.

3. Define the `edit_review` function.

4. Configure the `edit review` URL.

Updating movies.show template

In `/movies/templates/movies/show.html` file, add the following bold text:

```
...
{% for review in template_data.reviews %}
<li class="list-group-item pb-3 pt-3">
  <h5 class="card-title">
    Review by {{ review.user.username }}
  </h5>
  <h6 class="card-subtitle mb-2 text-muted">
    {{ review.date }}
  </h6>
  <p class="card-text">{{ review.comment }}</p>
  {% if user.is_authenticated and user ==
    review.user %}
  <a class="btn btn-primary"
    href="{% url 'movies.edit_review'
    id=template_data.movie.id
    review_id=review.id %}">Edit
  </a>
  {% endif %}
</li>
{% endfor %}
...
```

We added a code snippet for each review that is displayed. That code checks whether a user is authenticated and whether the user is the one who wrote a specific review. If both of these conditions are true, it will render the **Edit** button, which links to the `movies.edit_review` URL.

Creating the movies edit_review template

Now, in `/movies/templates/movies/`, create a new file, `edit_review.html`. For now, fill it in with the following:

```
{% extends 'base.html' %}
{% block content %}
<div class="p-3">
  <div class="container">
    <div class="row mt-3">
      <div class="col mx-auto mb-3">
```

```
        <h2>Edit Review</h2>
        <hr />
        <form method="POST">
          {% csrf_token %}
          <p>
            <label for="comment">Comment:</label>
            <textarea name="comment" required
            class="form-control" id="comment">{{
            template_data.review.comment }}</textarea>
          </p>
          <div class="text-start">
            <button type="submit"
              class="btn bg-dark text-white">Edit Review
            </button>
          </div>
        </form>
      </div>
    </div>
  </div>
</div>
{% endblock content %}
```

We have created a form to edit the review. This form is very similar to the review creation form. The differences are as follows:

- We removed the form action, which means that the form will be submitted to the current URL

- We displayed the current review comment value inside the text area

- We modified the button text

Defining the edit_review function

In /movies/views.py, add the following, as highlighted in bold:

```
from django.shortcuts import render, redirect, get_object_or_404

...

@login_required
def edit_review(request, id, review_id):
    review = get_object_or_404(Review, id=review_id)
    if request.user != review.user:
        return redirect('movies.show', id=id)
```

```
if request.method == 'GET':
    template_data = {}
    template_data['title'] = 'Edit Review'
    template_data['review'] = review
    return render(request, 'movies/edit_review.html',
        {'template_data': template_data})
elif request.method == 'POST' and
request.POST['comment'] != '':
    review = Review.objects.get(id=review_id)
    review.comment = request.POST['comment']
    review.save()
    return redirect('movies.show', id=id)
else:
    return redirect('movies.show', id=id)
```

Let's explain the preceding code:

- We import the get_object_or_404 function, which retrieves an object from the database or raises an HTTP 404 (Not Found) error (if the object is not found).

- We use the @login_required decorator to ensure that the edit_review function can only be accessed by authenticated users. If an unauthenticated user tries to access this function, they will be redirected to the login page.

- We define the edit_review function, which takes three parameters: the request, the movie ID, and the review ID.

- We retrieve the Review object with the given review_id. If the review does not exist, a 404 error will be raised.

- We check whether the current user (request.user) is the owner of the review to be edited (review.user). If the user does not own the review, the function redirects them to the movie.show page.

- Then, we check whether the request method is GET. In that case, the function prepares data for the template and renders the edit_review.html template.

- If the request method is POST and the comment field in the request's POST data is not empty, the function proceeds to update the review and redirects the user to the movie show page.

- In any other case, the function redirects the user to the movie show page.

> **Note**
>
> You can improve the look and feel of these functionalities by including your own error messages. You can use the login template and the login function, which uses and passes a template_data.error, as a base.

Configuring the edit_review URL

In /movies/urls.py, add the next path, as shown in bold:

```
from django.urls import path
from . import views

urlpatterns = [
    path('', views.index, name='movies.index'),
    path('<int:id>/', views.show, name='movies.show'),
    path('<int:id>/review/create/', views.create_review,
        name='movies.create_review'),
    path('<int:id>/review/<int:review_id>/edit/',
        views.edit_review, name='movies.edit_review'),
]
```

This path captures two integer values (the movie ID and review ID) from the URL and passes them to the edit_review function as arguments.

Now, save those files, run the server, and go to http://localhost:8000/movies. Click on a specific movie that contains a review you created, then click the **Edit** button (*Figure 9.6*).

Avatar

Description: A journey to a distant world and the battle for resources.

Price: $13

Reviews

Review by admin
April 11, 2024, 3:24 p.m.

Best movie ever!

Review by gara
April 11, 2024, 3:38 p.m.

Wow!

Figure 9.6 – A movie page with reviews and an edit button

An edit form will be shown. Modify the review and click the **Edit Review** button (*Figure 9.7*).

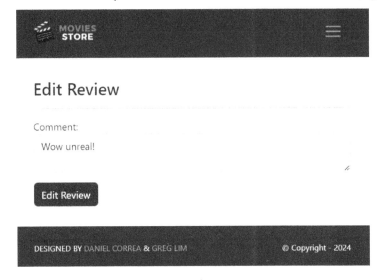

Figure 9.7 – The Edit Review page

You will be redirected to the movie show page. The new review comment should appear.

We just learned how to update reviews and models in general, so let's move to the final functionality and learn how to delete information.

Deleting a review

To be able to delete reviews, we need to follow the ensuing steps:

1. Update the `movies.show` template.
2. Define the `delete_review` function.
3. Configure the `delete review` URL.

Updating the movies.show template

In the `/movies/templates/movies/show.html` file, add the following bolded code:

```
...
<h5 class="card-title">
  Review by {{ review.user.username }}
</h5>
<h6 class="card-subtitle mb-2 text-muted">
```

```
                  {{ review.date }}
              </h6>
              <p class="card-text">{{ review.comment }}</p>
              {% if user.is_authenticated and user ==
                review.user %}
              <a class="btn btn-primary"
                href="{% url 'movies.edit_review'
                id=template_data.movie.id
                review_id=review.id %}">Edit
              </a>
              <a class="btn btn-danger"
                href="{% url 'movies.delete_review'
                id=template_data.movie.id
                review_id=review.id %}">Delete
              </a>
              {% endif %}
              ...
```

We have added a new delete button. This button links to the movies.delete_review URL, and much like the **Edit** button, it passes the movie ID and the review ID.

Defining the delete_review function

In /movies/views.py, add the following bold code at the end of the file:

```
    ...

@login_required
def delete_review(request, id, review_id):
    review = get_object_or_404(Review, id=review_id,
        user=request.user)
    review.delete()
    return redirect('movies.show', id=id)
```

Let's explain the preceding code:

- We use the @login_required decorator to ensure that the delete_review function can be only accessed by authenticated users. If an unauthenticated user tries to access this function, they will be redirected to the login page.

- We retrieve the Review object with the given review_id that belongs to the current user (request.user). If the review does not exist, or if the user does not own the review, an HTTP 404 error will be raised.

- We delete the review from the database using the Django model's delete() method.

- We redirect to the previous movie show page.

Configuring the delete_review URL

In /movies/urls.py, add the following path, as highlighted in bold:

```
...

urlpatterns = [
    ...
    path('<int:id>/review/<int:review_id>/edit/',
        views.edit_review, name='movies.edit_review'),
    path('<int:id>/review/<int:review_id>/delete/',
        views.delete_review, name='movies.delete_review'),
]
```

This path captures two integer values (the movie ID and the review ID) from the URL and passes them as arguments to the delete_review function.

Now, save those files, run the server, and go to http://localhost:8000/movies. Click on a specific movie that contains a review that you created, then click the **Delete** button (*Figure 9.8*).

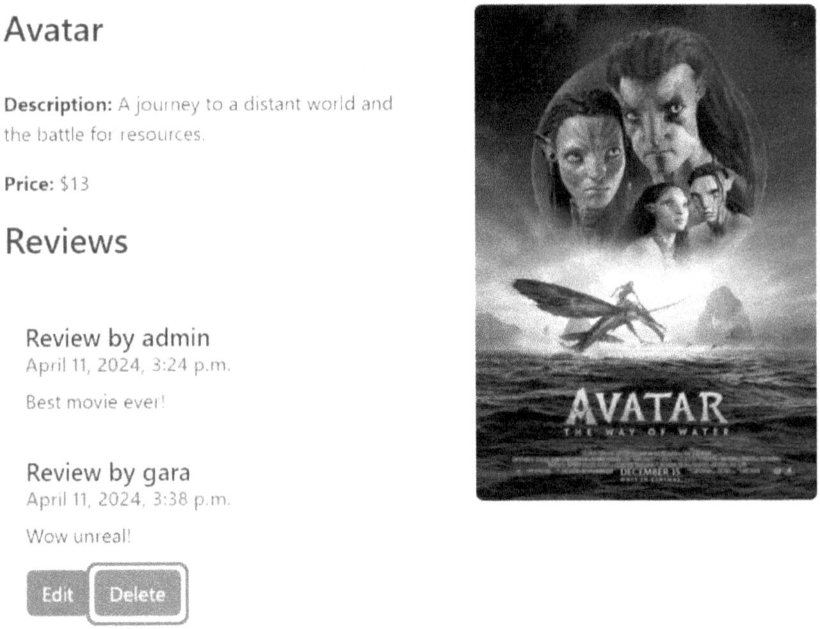

Figure 9.8 – A movie page with reviews and a Delete button

The review should be deleted, and you should be redirected to the movie show page.

Summary

In this chapter, we implemented a complete CRUD for movie reviews. With the tools we've developed, we can now create various CRUD systems by applying the knowledge gained in this chapter to other projects and models. As for the *Movies Store* project, users can now create, read, update, and delete reviews. Additionally, we have acquired the skills to manage application authorization, restricting access to certain routes and functions for non-logged-in users.

In the next chapter, we will learn how to create a shopping cart.

10

Implementing a Shopping Cart System

In this chapter, we'll learn all about how to make a shopping cart for websites. To implement this feature, we will need to learn how **web sessions** work and how to use **Django sessions**. Django sessions will be used to store user-specific information as they navigate through the site.

In this chapter, we will be covering the following topics:

- Introducing web sessions
- Creating a cart app
- Adding movies to the cart
- Listing movies added to the cart
- Removing movies from the cart

By the end of the chapter, you will have the knowledge to work with web sessions, implement shopping cart systems, and track and maintain user information between requests from the same user.

Technical requirements

In this chapter, we will be using **Python 3.10+**. Additionally, we will be using the **VS Code** editor in this book, which you can download from `https://code.visualstudio.com/`.

The code for this chapter is located at `https://github.com/PacktPublishing/Django-5-for-the-Impatient-Second-Edition/tree/main/Chapter10/moviesstore`.

The CiA video for this chapter can be found at `https://packt.link/mEcH8`

Introducing web sessions

Do you understand how the login system functions? How does the application recognize my connection status? How does it distinguish between displaying a logout button for a logged-in user and a login button for a friend who is not connected? How long does the application retain my connection status?

In this chapter, we'll address these questions and explore the significance of web sessions in the development of web applications. We will explore these elements in the following order:

1. HTTP protocol limitations
2.   Web sessions
3. Django login scenario
4. Django sessions

HTTP protocol limitations

Currently, our interaction with the Movies Store website relies on the HTTP protocol. For instance, to access movie information, we utilize `http://localhost:8000/movies`, and for logging in, we use `http://localhost:8000/login`. Each request we make utilizes the HTTP protocol as its communication medium.

Nevertheless, the HTTP protocol has its limitations. It operates in a **stateless** manner, implying that the server doesn't retain any information (state) between successive requests. With each new request, a fresh connection is established, removing any knowledge about previous interactions. Essentially, it lacks memory of past actions.

However, when logged into the application, subsequent requests display a logout button, indicating the application's ability to identify users and maintain state data. This functionality is achieved through Django sessions, which augment the capabilities of the HTTP protocol.

Web sessions

A **web session** comprises a sequence of continuous actions performed by a visitor on a website within a specified period. Each framework offers its own method for implementing sessions to monitor visitors' activities. *Figure 10.1* illustrates the functioning of Django sessions.

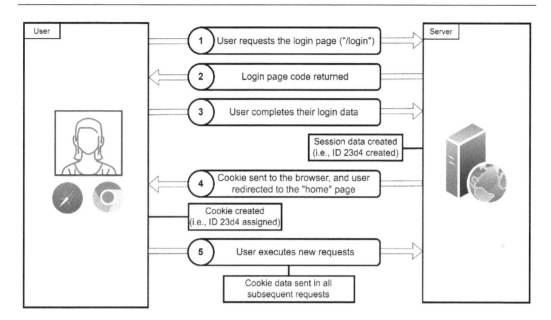

Figure 10.1 – An example of a Django session operation

Let's analyze the previous scenario:

1. The user navigates to the login page at `http://localhost:8000/login`.

2. Django then sends an HTTP response to the user, containing the login form.

3. The user fills out the login form and presses the **Login** button. Django validates the user data, and if it's accurate, creates session data for the user, and assigns them an ID or a **session key**. By default, the session data is stored in the database (we will see it in action in the next section).

4. Django sends a cookie to the user's browser. This cookie contains a session ID, which is used to retrieve the user's session data on subsequent requests. After successful login, Django redirects the user to the home page.

5. Each subsequent request will render the navigation menu with the **Logout** button, as the session ID is included in the user's requests. This process continues until the user clicks on the **Logout** button (which removes the session data and the cookie), or until the session expires, which defaults to two weeks.

Now that we have learned how Django sessions work, let's replicate it with our Movies Store project.

Django login scenario

Let's follow the next steps to see Django sessions in action:

1. Run the application in incognito mode, go to `http://localhost:8000/login`, and use the credentials of an already registered user to log in.

2. Go to `https://inloop.github.io/sqlite-viewer/`, drag and drop your `db.sqlite3` file onto the page, and then select the `django_session` table. You will see a new `session_key`, `session_data`, and `expire_date` corresponding to the user who just logged in (*Figure 10.2*).

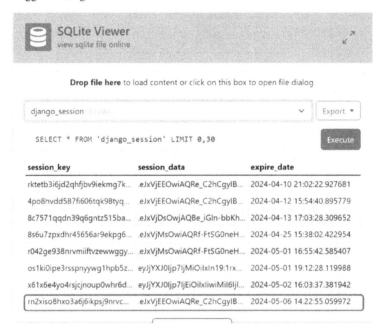

Figure 10.2 – The django_session table

3. The previous Django session data comes with a cookie. Go to your browser, verify that you're located on the **Movies Store** home page, and open the developer console. For Google Chrome, you can open the developer console with *Shift + Ctrl + J* (on Windows/Linux), or *option + ⌘ + J* (on macOS).

4. Then, navigate to the **Application** tab, click the **Cookies** | `http://127.0.0.1:8000` option, and you will see the stored cookie data, which includes a `sessionid` that matches with the `session_key` stored in the database (*Figure 10.3*).

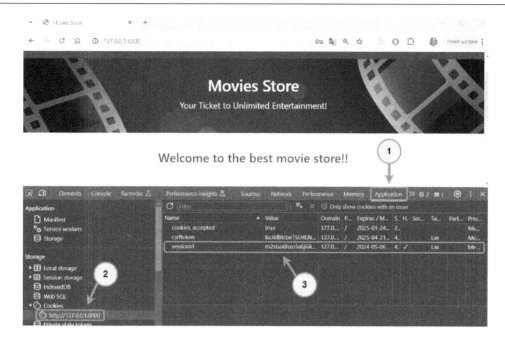

Figure 10.3 – The Movies Store cookie data

That's how Django tracks our website interactions; if you click the **Logout** button, all this data will disappear.

> **Note**
>
> Session data is not only created during the login scenario or exclusively for logged-in users. If your application utilizes Django's session functionalities at any point, it will also generate the corresponding session and cookie data. We will see this in action later when we implement the shopping cart system. You can find more information about Django sessions here: `https://docs.djangoproject.com/en/5.0/topics/http/sessions/`.

Django sessions

Django sessions are a mechanism for persisting data across HTTP requests in a web application. They allow Django to store and retrieve arbitrary data for a specific user across multiple requests. Here are some key points about Django sessions:

- **Client-side cookies**: By default, Django sessions are implemented using client-side cookies. Django commonly sets a unique session ID in the user's browser as a cookie.

- **Server-side storage**: While the session ID is stored in the client's browser, the actual session data is stored server-side. By default, the session data is stored in the database.

- **Configuration options**: Django offers various configuration options for sessions, including the session engine (e.g., database-backed, cached, file-based), session expiry time, and encryption of session data.

- **Integration with authentication**: Django sessions often work with Django's authentication system. For example, when a user logs in, their authentication status is typically stored in the session, allowing Django to keep the user logged in across multiple requests until they explicitly log out or the session expires.

- **Accessing session data**: Developers can access session data in Django views using the `request.session` attribute. This allows them to read, modify, and delete session data as needed during request processing.

Now that we've learned the fundamentals of web sessions and Django sessions, let's start creating the cart app.

Creating a cart app

All the shopping cart functionalities will be managed in their own application. So, let's create a cart app. Navigate to the top `moviesstore` folder (the one which contains the `manage.py` file) and run the following in the Terminal:

- For macOS, run the following command:

```
python3 manage.py startapp cart
```

- For Windows, run the following command:

```
python manage.py startapp cart
```

Figure 10.4 shows the new project structure. Verify it matches your current folder structure.

Figure 10.4 – The MOVIESSTORE project structure containing the cart app

Adding cart app in settings

Remember that for each newly created app, we must register it in the settings.py file. In /moviesstore/settings.py, under INSTALLED_APPS, add the following lines in bold:

```
...
INSTALLED_APPS = [
    ...
    'movies',
    'accounts',
    'cart',
]
...
```

Including the cart URL file in the project-level URL file

In /moviesstore/urls.py, add the following lines in bold:

```
...
urlpatterns = [
    path('admin/', admin.site.urls),
    path('', include('home.urls')),
    path('movies/', include('movies.urls')),
    path('accounts/', include('accounts.urls')),
    path('cart/', include('cart.urls')),
]
...
```

All the URLs that are defined in the cart.urls file will contain a cart/ prefix (as defined in the previous path). We will create the cart.urls file later.

Now that we have created the cart app, let's allow the addition of movies to the cart.

Adding movies to the cart

To allow the addition of movies to the cart, we will follow the next steps:

1. Configure the add_to_cart URL.
2. Define the add_to_cart function.
3. Update the movies.show template.

Configuring the add_to_cart URL

In /cart/, create a new file called urls.py. This file will contain the path regarding the URLs of the cart app. For now, fill it in with the following:

```
from django.urls import path
from . import views

urlpatterns = [
    path('<int:id>/add/', views.add, name='cart.add'),
]
```

We added a new path called cart/<int:id>/add (remember that the project-level URLs file defined a /cart prefix for this file). The <int:id> part indicates that this path expects an integer value to be passed from the URL and that the integer value will be associated with a variable named id. The id variable will be used to identify which movie we want to add to the cart. For example, the cart/1/add path indicates that we want to add the movie with id=1 to the cart.

Defining add_to_cart function

In /cart/views.py, add the following lines in bold:

```
from django.shortcuts import render
from django.shortcuts import get_object_or_404, redirect
from movies.models import Movie

def add(request, id):
    get_object_or_404(Movie, id=id)
    cart = request.session.get('cart', {})
    cart[id] = request.POST['quantity']
    request.session['cart'] = cart
    return redirect('home.index')
```

Let's explain the previous code:

- We import the redirect and get_object_or_404 functions. We also import the Movie model from the "movies" app.

- We define the add function, which takes two parameters: the request and the movie ID.

- We fetch the Movie object with the given id from the database (by using the get_object_or_404 function). If no such object is found, a 404 (Not Found) error is raised.

- We check the session storage for a key called 'cart'. If the key does not exist, a { } empty dictionary is assigned to the cart variable.

- We modify the `cart` variable. We add a new key to the `cart` dictionary based on the movie ID, and the corresponding value based on the movie quantity the user wants to add to the cart (we will collect `quantity` through an HTML form later). For example, if the user wants to add 2 movies with `id=1`, a new key/value such as this `cart["1"] = "2"` will be added to the dictionary.

- The updated `cart` dictionary is then saved back to the session using `request.session['cart'] = cart`.

- After updating the cart, we redirect the user to the home page (`home.index`).

Now, let's update the `movies.show` template to include a form to add movies to the cart.

Updating the movies.show template

In the `/movies/templates/movies/show.html` file, add the following lines in bold:

```
...
<p><b>Description:</b> {{
  template_data.movie.description }}</p>
<p>
  <b>Price:</b> ${{ template_data.movie.price }}
</p>
<p class="card-text">
  <form method="post"
    action="{% url 'cart.add'
    id=template_data.movie.id %}">
    <div class="row">
      {% csrf_token %}
      <div class="col-auto">
        <div class="input-group col-auto">
          <div class="input-group-text">Quantity
            </div>
          <input type="number" min="1" max="10"
            class="form-control quantity-input"
            name="quantity" value="1">
        </div>
      </div>
      <div class="col-auto">
        <button class="btn bg-dark text-white"
          type="submit">Add to cart</button>
      </div>
    </div>
  </form>
</p>
...
```

We added a new form to allow users to add movies to the cart. This form also includes an input field to specify the quantity of the movie the user wishes to add to the cart. The form is linked to the `'cart.add'` path and passes the movie `id` as part of the form action.

Now, save those files, run the server, go to `http://localhost:8000/movies`, click on a specific movie, and you will see the **Add to cart** functionality available (*Figure 10.5*).

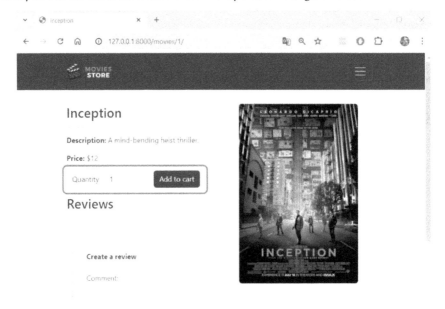

Figure 10.5 – Movie page with the Add to cart functionality

We successfully added a button to add movies to the cart. Now, let's move on to listing the movies added to the cart.

Listing movies added to the cart

To allow us to list movies added to the cart, we will follow the next steps:

1. Configure the cart index URL.
2. Define a `utils` file.
3. Define a filter.
4. Define an `index` function.
5. Creating the `cart.index` template.
6. Updating the `add_to_cart` function.
7. Adding a link in the base template.

Configuring cart index URL

In /cart/urls.py, add the next path by adding the lines in bold:

```
from django.urls import path
from . import views

urlpatterns = [
    path('', views.index, name='cart.index'),
    path('<int:id>/add/', views.add, name='cart.add'),
]
```

We defined a ' ' path but remember that the project-level URL file defined a /cart prefix for this file. So, if a URL matches the /cart path, it will execute the index function defined in the views file. We will implement the index function later.

Defining a utils file

Commonly, cart pages display the total amount to be paid. This is calculated by summing the prices of the movies based on their corresponding quantities. We will define a calculate_cart_total function to perform this process. However, it is not recommended to place this kind of function in either the views file or the models file. Therefore, we will create a utilities (utils) file in which we will place this function to be easily reused.

In /cart/, create a new file called utils.py. For now, fill it in with the following:

```
def calculate_cart_total(cart, movies_in_cart):
    total = 0
    for movie in movies_in_cart:
        quantity = cart[str(movie.id)]
        total += movie.price * int(quantity)
    return total
```

Let's explain the previous code:

- We define the calculate_cart_total function, which takes two parameters: cart and movies_in_cart. The cart parameter is a dictionary that represents the user's shopping cart. Remember that the keys are strings representing movie IDs, and the values are strings representing the quantities of each movie in the cart. The movies_in_cart parameter is a list of Movie objects representing the movies in the cart.

- We initialize a total variable by 0.

- We iterate through the list of movies in the cart. For each movie, we extract the corresponding quantity added to the cart and multiply it by the movie's price. Then, we add the total cost for the movie to the total variable.

- Finally, we return the total variable.

In summary, `calculate_cart_total` calculates the total cost of the movies in the user's cart by iterating over each movie in the cart, multiplying the movie's price by its quantity, and summing up the total costs. This function will be used later in the `views` file.

Defining a filter

Django filters are a feature of the template engine that allows you to modify or format data in a template. Filters are applied to variables using a pipe (|) character and are a powerful tool for customizing the presentation of data in a template.

We want to list the movies added to the cart and display the quantity of each movie. This requires accessing the cart session data and using each movie's ID as a key to get the `quantity` value. While this may not sound very complicated, Django templates are designed to be simple and primarily focused on rendering data provided by views. Sometimes, Django templates don't allow access to data that depends on other variables or complex data structures. In such cases, you will need to create a custom filter or tag. We will create a custom filter to access the `quantity` data for the movies in the cart.

In `/cart/`, create a `templatetags` folder. Then, in `/cart/templatetags/` create a new file called `cart_filters.py`. For now, fill it in with the following:

```
from django import template

register = template.Library()

@register.filter(name='get_quantity')
def get_cart_quantity(cart, movie_id):
    return cart[str(movie_id)]
```

Let's explain the previous code:

- We import the `template` module, which provides utilities for working with Django templates.

- We use the `register = template.Library()` code to create an instance of `template.Library`, which is used to register custom template tags and filters.

- We use the `@register.filter(name='get_quantity')` decorator to register the `get_cart_quantity` function as a custom template filter named `get_quantity`. The `name='get_quantity'` argument specifies the name of the filter as it will be used in templates.

- We define the `get_cart_quantity` function, which takes two arguments: the `cart` session dictionary, and the ID of the movie for which the quantity is needed.

- We access the `quantity` value by using the `cart` dictionary and `movie_id` as the key. We convert `movie_id` to a string to ensure compatibility with the cart keys.

- Finally, we return the corresponding `quantity` value.

> **Note**
>
> Once the custom filter is defined and registered, you can use it in a Django template such as this: `{{ request.session.cart|get_quantity:movie.id }}`
>
> In the preceding example, the `get_quantity` filter is applied to `request.session.cart` with the `movie.id` argument to obtain the quantity of the specific movie in the cart.
>
> You can find more information about custom filters and tags here: `https://docs.djangoproject.com/en/5.0/howto/custom-template-tags/`.

We have designed all the elements we need to implement the cart index function.

Defining an index function

In `/cart/views.py`, add the following in bold:

```
from django.shortcuts import render
from django.shortcuts import get_object_or_404, redirect
from movies.models import Movie
from .utils import calculate_cart_total

def index(request):
    cart_total = 0
    movies_in_cart = []
    cart = request.session.get('cart', {})
    movie_ids = list(cart.keys())
    if (movie_ids != []):
        movies_in_cart =
          Movie.objects.filter(id__in=movie_ids)
        cart_total = calculate_cart_total(cart,
            movies_in_cart)

    template_data = {}
    template_data['title'] = 'Cart'
    template_data['movies_in_cart'] = movies_in_cart
    template_data['cart_total'] = cart_total
    return render(request, 'cart/index.html',
        {'template_data': template_data})

def add(request, id):
    ...
```

Let's explain the previous code:

- We import the `calculate_cart_total` function from the `utils` file.

- We define the `index` function.

- We initialize the `cart_total` to 0, and `movies_in_cart` as an empty list.

- We retrieve the cart information from the session using `request.session.get('cart', {})`.

- We extract the movie IDs that were added to the cart based on the cart keys.

- If there are any movie IDs in the cart, the function queries the database for movies with those IDs using `Movie.objects.filter(id__in=movie_ids)`. Additionally, we calculate the total cost of the movies in the cart using the `calculate_cart_total` function, which updates the `cart_total` variable.

- Finally, we prepare the `template_data` dictionary and render the `cart/index.html` template.

In summary, the `index` function is designed to render the cart page, showing the movies in the cart and the total cost of those movies.

Creating the cart.index template

In `/cart/`, create a `templates` folder. Then, in `/cart/templates/`, create a `cart` folder.

Now, in `/cart/templates/cart/`, create a new file, `index.html`. For now, fill it in with the following:

```
{% extends 'base.html' %}
{% block content %}
{% load static %}
{% load cart_filters %}
<div class="p-3">
  <div class="container">
    <div class="row mt-3">
      <div class="col mx-auto mb-3">
        <h2>Shopping Cart</h2>
        <hr />
      </div>
    </div>
    <div class="row m-1">
      <table class="table table-bordered table-striped
        text-center">
        <thead>
```

```
            <tr>
              <th scope="col">ID</th>
              <th scope="col">Name</th>
              <th scope="col">Price</th>
              <th scope="col">Quantity</th>
            </tr>
          </thead>
          <tbody>
            {% for movie in template_data.movies_in_cart %}
            <tr>
              <td>{{ movie.id }}</td>
              <td>{{ movie.name }}</td>
              <td>${{ movie.price }}</td>
              <td>{{
                 request.session.cart|get_quantity:movie.id }}
              </td>
            </tr>
            {% endfor %}
          </tbody>
        </table>
      </div>
      <div class="row">
        <div class="text-end">
          <a class="btn btn-outline-secondary mb-2"><b>Total
            to pay:</b> ${{ template_data.cart_total }}</a>
        </div>
      </div>
    </div>
  </div>
{% endblock content %}
```

Let's explain the previous code:

- We use the {% load cart_filters %} tag, which loads custom template filters defined in the cart_filters file. In our case, it includes the filter named get_quantity.

- We create an HTML table.

- We iterate through the movies in the cart. For each movie, we display its id, name, price, and quantity value in the cart. To display the quantity in the cart, we use the get_quantity filter.

- Finally, we display the cart_total value.

Updating the add_to_cart function

In /cart/views.py, add the following lines in bold:

```
...
def add_to_cart(request, id):
    get_object_or_404(Movie, id=id)
    cart = request.session.get('cart', {})
    cart[id] = request.POST['quantity']
    request.session['cart'] = cart
    return redirect('cart.index')
```

If a user adds a movie to the cart, they will now be redirected to the cart page.

Adding a link in the base template

Finally, let's add the cart link in the base template. In /moviesstore/templates/base.html, in the header section, add the following lines in bold:

```
...
<a class="nav-link"
  href="{% url 'home.about' %}">About</a>
<a class="nav-link"
  href="{% url 'movies.index' %}">Movies</a>
<a class="nav-link"
  href="{% url 'cart.index' %}">Cart</a>
...
```

Now, save those files, run the server, go to http://localhost:8000/movies, click on a couple of movies, and add them to the cart. Then, go to the **Cart** section and you will see a **Cart** page with its corresponding information (*Figure 10.6*).

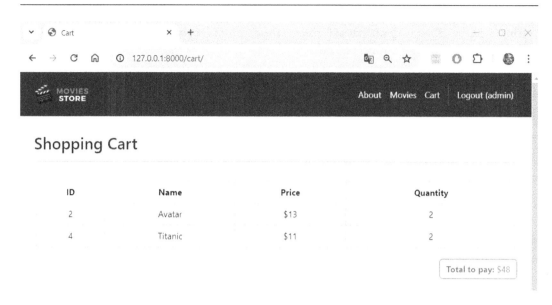

Figure 10.6 – The Cart page

Now that we have learned how to implement a cart system, let's finish this chapter by including the functionality to remove movies from the cart.

Removing movies from the cart

To remove movies from the cart, we will follow the next steps:

1. Configure a `clear` URL.

2. Defining a `clear` function.

3. Updating the `cart.index` template.

Configuring clear URL

In `/cart/urls.py`, add the next path by adding the following lines in bold:

```
from django.urls import path
from . import views

urlpatterns = [
    path('', views.index, name='cart.index'),
    path('<int:id>/add/', views.add, name='cart.add'),
    path('clear/', views.clear, name='cart.clear'),
]
```

We define a `cart/clear/` path that will execute the `clear` function defined in the `views` file. We will implement the `clear` function later.

Defining clear function

In `/cart/views.py`, add the following lines in bold at the end of the file:

```
...

def clear(request):
    request.session['cart'] = {}
    return redirect('cart.index')
```

We just update the user's cart session to an empty dictionary. This removes all previous movies added to the cart, and we redirect the user to the cart page.

Updating the cart.index template

In the `/cart/templates/cart/index.html` file, add the following lines in bold:

```
...
<div class="row">
  <div class="text-end">
    <a class="btn btn-outline-secondary mb-2"><b>Total
      to pay:</b> ${{ template_data.cart_total }}</a>
    {% if template_data.movies_in_cart|length > 0 %}
    <a href="{% url 'cart.clear' %}">
      <button class="btn btn-danger mb-2">
        Remove all movies from Cart
      </button>
    </a>
    {% endif %}
  </div>
</div>
...
```

We added an `if` section to check whether there are any movies in the cart. If there are, we display a button that allows the user to clear the cart.

Now, save those files, run the server, go to `http://localhost:8000/movies`, click on a couple of movies, and add them to the cart. Then, go to the **Cart** section and click **Remove all movies from Cart** (*Figure 10.7*). You will see an empty cart.

Shopping Cart

ID	Name	Price	Quantity
1	Inception	$12	2
2	Avatar	$13	2

Total to pay: $50   Remove all movies from Cart

Figure 10.7 – The Cart page

Summary

In this chapter, we learned how web sessions and Django sessions work. We created a cart app that allows us to add movies to the cart, list the movies added to the cart, and remove the movies from the cart. We also learned that the `utils` file is useful for storing functions that can be reused across our app. Additionally, we learned that filters allow us to modify or format the data displayed in the templates, and we learned how to utilize some Django session functionalities. In the next chapter, we will create the order and item models to enable users to purchase movies.

11

Implementing Order and Item Models

In the previous chapter, we implemented the shopping cart system and allowed users to add movies to the cart. To enable users to purchase movies, we need to store additional information in the database, specifically to store order and item information. In this chapter, we will implement the order and item models and establish a connection between them.

In this chapter, we will be covering the following topics:

- Analyzing store invoices
- Creating the order model
- Creating the item model
- Recapping the Movies Store class diagram

By the end of the chapter, we will have the complete structure for storing purchase information. Additionally, we will recap the class diagram and examine the relationship between the Django models and the classes in the class diagram.

Technical requirements

In this chapter, we will be using **Python 3.10+**. Additionally, we will be using the **VS Code** editor in this book, which you can download from `https://code.visualstudio.com/`.

The code for this chapter is located at `https://github.com/PacktPublishing/Django-5-for-the-Impatient-Second-Edition/tree/main/Chapter11/moviesstore`.

The CiA video for this chapter can be found at `https://packt.link/eQzNG`

Analyzing store invoices

If you purchase something in a modern store, it is almost certain that you will receive an invoice. Different stores manage invoices with varying information, but in most cases, you will find the same essential information. *Figure 11.1* shows a simple invoice. We will use this as a blueprint to design and implement the Django models that we will use to store the purchase information.

Order #1

Date: 2024-04-22

Total: $50

User ID - username: 1 - daniel

ID	Movie ID - name	Qty	Price
1	1 - Inception	2	12
2	2 - Avatar	2	13

Figure 11.1 – Example of a simple invoice

Let's analyze the invoice shown in *Figure 11.1* to understand the kind of information we need to store for purchase (based on orders and items).

We must store the following information for the order:

- **ID**: To uniquely identify each order. In the previous figure, it is represented by **#1**.

- **Date**: To identify the date on which the order was completed. In the previous figure, it is represented by **2024-04-22**.

- **Total**: To identify the total amount of the order. In the previous figure, it is represented by **$50**.

- **User**: To identify the user who made the purchase. In the previous figure, it is represented by **1 - daniel**.

An order is composed of items, represented as the internal table in *Figure 11.1*. We must store the following information for each item:

- **ID**: To uniquely identify each item. In the previous figure, the ID of the first item is represented by **1**.

- **Quantity**: To specify the quantity of the movie the user wants to purchase. In the previous example, the quantity of the first item is represented by **2**.

- **Price**: To specify the price of the movie at which the user purchased the item. In the previous example, the price of the first item is represented by **12**.

- **Movie**: To specify the movie to which the item is linked. In the previous example, the linked movie of the first item is represented by **1 - Inception**.

- **Order**: To specify the order to which the item is linked. In the previous example, the linked order of the first item is represented by **#1**.

Now that we have grasped the functioning of these simple invoices, let's proceed to create the appropriate models.

Creating the order model

To store the purchase information, we need to start by creating an Order Django model. The following are the three steps for storing the purchase information:

1. Create the Order model.

2. Apply migrations.

3. Add the order model to the admin panel.

Let's go through them in detail.

Creating the Order model

We will start creating the Order model. We will create this model inside the cart app.

In /cart/models.py file, add the following in *bold*:

```
from django.db import models
from django.contrib.auth.models import User

class Order(models.Model):
    id = models.AutoField(primary_key=True)
    total = models.IntegerField()
    date = models.DateTimeField(auto_now_add=True)
    user = models.ForeignKey(User,
        on_delete=models.CASCADE)

    def __str__(self):
        return str(self.id) + ' - ' + self.user.username
```

Let's explain the previous code:

- We import the `User` model from Django's `django.contrib.auth.models` module.

- We define a Python class named `Order`, which inherits from `models.Model`. This means that `Order` is a Django model class.

- Inside the `Order` class, we define several fields:

  - `id`: This is an `AutoField`, which automatically increments its value for each new record added to the database. The `primary_key=True` parameter specifies that this field is the primary key for the table, uniquely identifying each record.

  - `total`: This is a `IntegerField`, which represents the total amount of the order. It stores integer values.

  - `date`: This is a `DateTimeField`, which represents the date and time when the order was created. `auto_now_add=True` ensures that the date and time are automatically set to the current date and time when the order is created.

  - `user`: This is a foreign key relationship to the `User` model, which establishes a many-to-one relationship between orders and users. It means that each order is associated with a single user, and each user can have multiple orders. `on_delete=models.CASCADE` specifies that if the related user is deleted, the associated orders will also be deleted.

- `__str__` is a method that returns a string representation of the order. In this case, it returns a string composed of the order ID and the username of the user who placed the order.

Applying migrations

Now that we have created the `Order` model, let's update our database by running one of the following commands, depending on your operating system:

- For macOS, run this:

```
python3 manage.py makemigrations
python3 manage.py migrate
```

- For Windows, run this:

```
python manage.py makemigrations
python manage.py migrate
```

Now, you should see something like this:

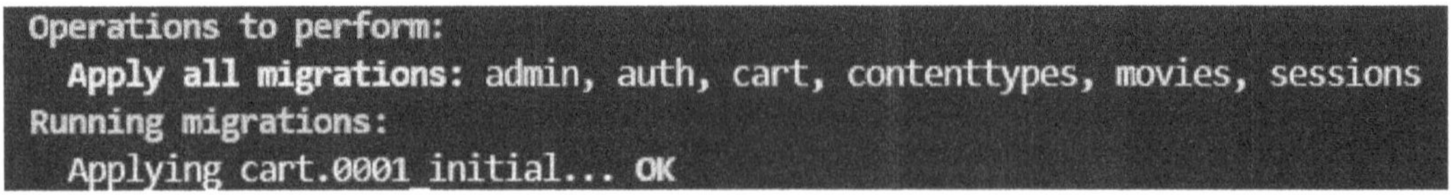

```
Operations to perform:
  Apply all migrations: admin, auth, cart, contenttypes, movies, sessions
Running migrations:
  Applying cart.0001_initial... OK
```

Figure 11.2 – Applying the order migration

Adding the order model to the admin panel

To add the Order model to admin, go to /cart/admin.py and register it by adding the following in *bold*:

```
from django.contrib import admin
from .models import Order


admin.site.register(Order)
```

When you save your file, stop the server, run the server, and go back to /admin. The order model will now appear (as shown in *Figure 11.3*):

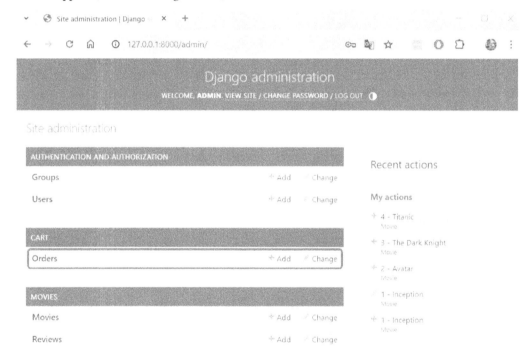

Figure 11.3 – Admin page with orders available

Now that we have created and applied our Order model, let's create the Item model to complete the information required to store purchases.

Creating the Item model

Let's continue by creating an Item model and follow these steps:

1. Create the Item model.

2. Apply migrations.

3. Add the item model to the admin panel.

Creating the Item model

In /cart/models.py file, add the following in *bold*:

```
from django.db import models
from django.contrib.auth.models import User
from movies.models import Movie

class Order(models.Model):
    ...


class Item(models.Model):
    id = models.AutoField(primary_key=True)
    price = models.IntegerField()
    quantity = models.IntegerField()
    order = models.ForeignKey(Order,
        on_delete=models.CASCADE)
    movie = models.ForeignKey(Movie,
        on_delete=models.CASCADE)

    def __str__(self):
        return str(self.id) + ' - ' + self.movie.name
```

Let's explain the previous code:

- We import the Movie model from the movies app.

- We define a Python class named Item, which inherits from models.Model. This means that Item is a Django model class.

- Inside the Item class, we define several fields:

  - id: This is an AutoField, which automatically increments its value for each new record added to the database. The primary_key=True parameter specifies that this field is the primary key for the table, uniquely identifying each record.

  - price: This is an IntegerField, which represents the price at which the item was purchased.

- quantity: This is an `IntegerField`, which represents the desired quantity of the item to purchase.

- order: This is a foreign key relationship with the `Order` model, which defines a foreign key relating each item to a specific order.

- movie: This is a foreign key relationship with the `Movie` model, which defines a foreign key relating each item to a specific movie.

- `__str__` is a method that returns a string representation of the item. In this case, it returns a string composed of the item ID and the name of the associated movie.

Applying migrations

Now that we have created the `Item` model, let's update our database by running the following commands based on your operating system.

- For macOS, run this:

```
python3 manage.py makemigrations
python3 manage.py migrate
```

- For Windows, run this:

```
python manage.py makemigrations
python manage.py migrate
```

Now, you should see something like this:

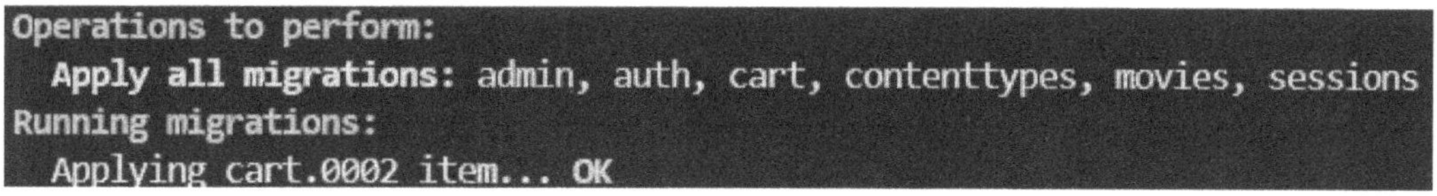

```
Operations to perform:
  Apply all migrations: admin, auth, cart, contenttypes, movies, sessions
Running migrations:
  Applying cart.0002 item... OK
```

Figure 11.4 – Applying the item migration

Adding the item model to the admin panel

To add the `Item` model to admin, go to `/cart/admin.py` and register it by adding the following in *bold*:

```
from django.contrib import admin
from .models import Order, Item

admin.site.register(Order)
admin.site.register(Item)
```

After saving your file, stop the server and then run the server again. Then, go back to /admin. The item model will now appear (as shown in *Figure 11.5*):

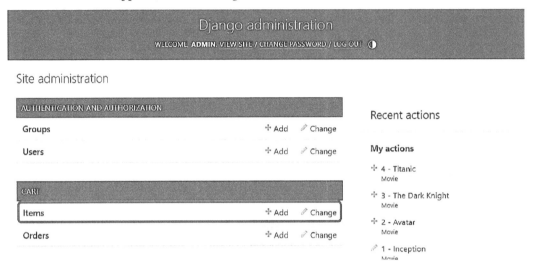

Figure 11.5 – Admin page with items available

Now we have completed the data structure required to make purchases. But before proceeding with the purchase process, let's recap how our models relate to the project's class diagram.

Recapping the Movies Store class diagram

The class diagram of the Movies Store that we designed in *Chapter 1* served as a blueprint for designing the code of the Movies Store. We have already implemented all the models required to complete the project code. So, let's quickly recap this relationship between models and classes.

Figure 11.6 shows the class diagram, highlighting the locations where we implemented the corresponding Django models:

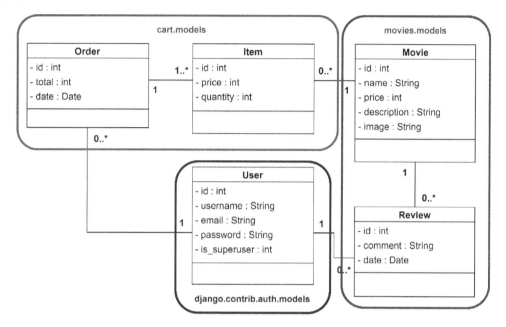

Figure 11.6 – Movies Store class diagram, highlighting model locations

Let's analyze the previous figure:

- The Movie and Review models were implemented inside the movies app.

- The Order and Item models were implemented inside the cart app.

- The User model was not implemented. Instead, we took advantage of the provided Django built-in model located in the admin.contrib.auth app.

Finally, let's review how a specific class relates to a model (*Figure 11.7*):

- **1**: The Review class name became a Review Python class. We inherited from models.Model to define it as a Django model class.

- **2**: The id, comment, and date class attributes became Python class attributes. We utilized the models module to utilize available field types similar to those defined in the class diagram.

- **3**: The relationship between the Review and Movie classes became a Python class attribute. We utilized the models.ForeignKey method to define a foreign key relationship between the two models.

- **4:** The relationship between the `Review` and `User` classes became a Python class attribute. We utilized the `models.ForeignKey` method to define a foreign key relationship between the two models.

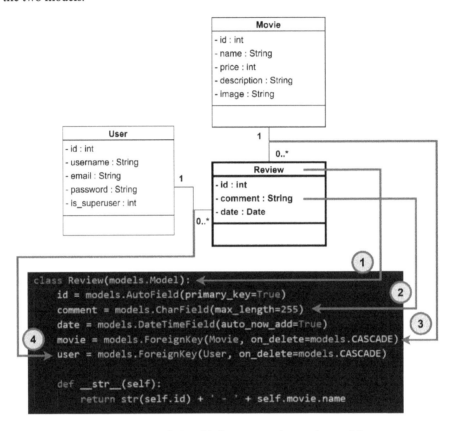

Figure 11.7 – Relationship between a class and a model

We have completed all the connections between the class diagram and the Django models. Now, we are ready to enable users to make purchases.

Summary

In this chapter, we learned how simple invoices work. We created a couple of models (`Order` and `Item`). These models will allow us to store information about the users' purchases. We recapitulated the process of creating Django models and applying migrations. In the end, we reviewed how class diagrams served as a blueprint to create our project models. In the next chapter, we will implement purchase functionality and allow users to view their orders.

12

Implementing the Purchase and Orders Pages

During the previous chapter, we implemented the models required to store the purchase information. In this chapter, we will implement the purchase functionality and finalize the Movies Store project with an orders page. Users will be able to check their placed orders. Later, we will recap the Movies Store MVT architecture to check the consistency between the Python code and the architecture diagram.

In this chapter, we will be covering the following topics:

- Creating the purchase page
- Creating the orders page
- Recapping the Movies Store MVT architecture

By the end of the chapter, we will have the complete code for our Movies Store project. We will also be capable of relating architecture diagrams to the actual implemented code.

Technical requirements

In this chapter, we will be using Python 3.10+. Additionally, we will be using the **VS Code** editor in this book, which you can download from `https://code.visualstudio.com/`.

The code for this chapter is located at `https://github.com/PacktPublishing/Django-5-for-the-Impatient-Second-Edition/tree/main/Chapter12/moviesstore`.

The CiA video for this chapter can be found at `https://packt.link/4NyAv`

Creating the purchase page

Let's improve our shopping cart page and include some functionalities to allow users to make purchases. To achieve that, we need to follow these steps:

1. Configuring the purchase URL.

2. Defining the purchase function.

3. Updating the cart.index template.

4. Creating the cart.purchase template.

Configuring the purchase URL

In /cart/urls.py, add the next path as shown in **bold**:

```
from django.urls import path
from . import views

urlpatterns = [
    path('', views.index, name='cart.index'),
    path('<int:id>/add/', views.add, name='cart.add'),
    path('clear/', views.clear, name='cart.clear'),
    path('purchase/', views.purchase,
        name='cart.purchase'),
]
```

We defined a cart/purchase/ path that will execute the purchase function defined in the views file. We will implement the purchase function later.

Defining the purchase function

In /cart/views.py, add the following lines of code in **bold**:

```
...
from movies.models import Movie
from .utils import calculate_cart_total
from .models import Order, Item
from django.contrib.auth.decorators import login_required


...


@login_required
def purchase(request):
    cart = request.session.get('cart', {})
```

```
    movie_ids = list(cart.keys())

    if (movie_ids == []):
        return redirect('cart.index')

    movies_in_cart = Movie.objects.filter(id__in=movie_ids)
    cart_total = calculate_cart_total(cart, movies_in_cart)

    order = Order()
    order.user = request.user
    order.total = cart_total
    order.save()

    for movie in movies_in_cart:
        item = Item()
        item.movie = movie
        item.price = movie.price
        item.order = order
        item.quantity = cart[str(movie.id)]
        item.save()

    request.session['cart'] = {}
    template_data = {}
    template_data['title'] = 'Purchase confirmation'
    template_data['order_id'] = order.id
    return render(request, 'cart/purchase.html',
        {'template_data': template_data})
```

The previous function is the largest one we have implemented in this book. Let's explain this function by breaking it down into parts:

- `from .models import Order, Item`

  `from django.contrib.auth.decorators import login_required`

  Let's analyze this piece of code:

  - We import the `Order` and `Item` models from the current app directory.

  - We import the `login_required` decorator.

- `@login_required`

  `def purchase(request):`

  `    cart = request.session.get('cart', {})`

  `    movie_ids = list(cart.keys())`

```
if (movie_ids == []):
    return redirect('cart.index')
```

Let's analyze this piece of code:

- We use the `login_required` decorator to ensure that the user must be logged in to access the `purchase` function.

- We define the `purchase` function, which will handle the purchase process.

- We retrieve the cart data from the user's session. The `cart` variable will contain a dictionary with movie IDs as keys and quantities as values.

- We retrieve the movie IDs stored in the `cart` dict and convert them into a list named `movie_ids`.

- We check if the `movie_ids` list is empty (which indicates the cart is empty). In this case, the user is redirected to the `cart.index` page (here, the `purchase` function finalizes its execution).

-
  ```
  movies_in_cart = Movie.objects.filter(id__in=movie_ids)
  cart_total = calculate_cart_total(cart, movies_in_cart)

  order = Order()
  order.user = request.user
  order.total = cart_total
  order.save()

  for movie in movies_in_cart:
      item = Item()
      item.movie = movie
      item.price = movie.price
      item.order = order
      item.quantity = cart[str(movie.id)]
      item.save()
  ```

Let's analyze this piece of code:

- If the cart is not empty, we continue the purchase process.

- We retrieve movie objects from the database based on the IDs stored in the cart using `Movie.objects.filter(id__in=movie_ids`.

- We calculate the total cost of the movies in the cart using the `calculate_cart_total()` function.

- We create a new `Order` object. We set its attributes such as `user` (the logged-in user) and `total` (the cart total), and save it to the database.

- We iterate over the movies in the cart. We create an `Item` object for each movie in the cart. For each `Item`, we set its `price` and `quantity`, link the corresponding `movie` and `order`, and save it to the database.

-
```
request.session['cart'] = {}
template_data = {}
template_data['title'] = 'Purchase confirmation'
template_data['order_id'] = order.id
return render(request, 'cart/purchase.html',
    {'template_data': template_data})
```

Let's analyze this piece of code:

- After the purchase is completed, we clear the cart in the user's session by setting `request.session['cart']` to an empty dictionary.

- We prepare the data to be sent to the purchase confirmation template. This data includes the title of the page and the ID of the created order.

- Finally, we render the `cart/purchase.html` template.

Now that we have finished the purchase function, let's include a button that links to this function.

Updating cart.index template

In the `/cart/templates/cart/index.html` file, add the following lines in **bold**:

```
...
<a class="btn btn-outline-secondary mb-2">
  <b>Total to pay:</b> ${{ template_data.cart_total
}}</a>

{% if template_data.movies_in_cart|length > 0 %}
<a href="{% url 'cart.purchase' %}"
```

```
        class="btn bg-dark text-white mb-2">Purchase
      </a>
      <a href="{% url 'cart.clear' %}">
        <button class="btn btn-danger mb-2">
          Remove all movies from Cart
        </button>
      </a>
      {% endif %}
      ...
```

We have added a button that links the shopping cart page with the purchase function. This button will be only displayed if there are movies added to the cart.

Creating cart.purchase template

Now, in `/cart/templates/cart/`, create a new file, `purchase.html`. For now, fill it with the following:

```
{% extends 'base.html' %}
{% block content %}
<div class="p-3">
  <div class="container">
    <div class="row mt-3">
      <div class="col mx-auto mb-3">
        <h2>Purchase Completed</h2>
        <hr />
        <p>Congratulations, purchase completed. Order
          number is: <b>#{{ template_data.order_id }}</b>
        </p>
      </div>
    </div>
  </div>
</div>
{% endblock content %}
```

We have created a simple template that extends the `base.html` template and shows a congratulations message to the user, including the order number of the current purchase.

Now, save those files, run the server, go to `http://localhost:8000/movies`, click on a couple of movies, and add them to the cart. Then, go to the **Cart** section and click **Purchase** (you will need to be logged in to execute the purchase action). Then, you will see a purchase confirmation message (*Figure 12.1*):

Figure 12.1 – Purchase page

If you navigate to the admin panel, you will see a new order registered (linked to the user who made the purchase) and a couple of items (linked to the previous order), as shown in *Figure 12.2*:

Figure 12.2 – Order and items in the admin panel

At this point, we are able to create orders and register the corresponding information into the database. Now, let's implement a page to see the orders.

Creating the orders page

Let's finalize our Movies Store by allowing users to see their orders. To achieve that, we need to follow these steps:

1. Configuring the orders URL.
2. Defining the `orders` function.
3. Creating the `accounts.orders` template.
4. Adding a link to the base template.

Configuring the orders URL

An order belongs to a specific user. Because of this, we will add the orders functionality inside the `accounts` app. In `/accounts/urls.py`, add the next path in **bold**:

```
from django.urls import path
from . import views

urlpatterns = [
    path('signup', views.signup, name='accounts.signup'),
    path('login/', views.login, name='accounts.login'),
    path('logout/', views.logout, name='accounts.logout'),
    path('orders/', views.orders, name='accounts.orders'),
]
```

We defined an `accounts/orders/` path, which will execute the `orders` function defined in the `views` file. We will implement the `orders` function later.

Defining the orders function

In `/accounts/views.py`, add the following lines in **bold**:

```
...
from django.shortcuts import redirect
from django.contrib.auth.decorators import login_required
from django.contrib.auth.models import User

...

@login_required
def orders(request):
    template_data = {}
    template_data['title'] = 'Orders'
    template_data['orders'] = request.user.order_set.all()
    return render(request, 'accounts/orders.html',
        {'template_data': template_data})
```

Let's explain the previous code:

- We import the `User` model from Django's authentication system.

- We use the `login_required` decorator to ensure that the user must be logged in to access the `orders` function.

- We define the `orders` function, which takes a `request` object as a parameter.

- We define the `template_data` variable and assign it a `title`.

- We retrieve all orders belonging to the currently logged-in user (`request.user`). The `order_set` attribute is used to access the related orders associated with the user through their relationship (you can learn more about this type of relationship here `https://docs.djangoproject.com/en/5.0/topics/db/examples/many_to_one/`). Remember that there is a `ForeignKey` relationship between the `User` model and the `Order` model.

- Finally, we pass the orders to the template and render it.

Creating accounts.orders template

Now, in `/accounts/templates/accounts/`, create a new file, `orders.html`. For now, fill it with the following:

```
{% extends 'base.html' %}
{% block content %}
<div class="p-3">
  <div class="container">
    <div class="row mt-3">
      <div class="col mx-auto mb-3">
        <h2>My Orders</h2>
        <hr />
        {% for order in template_data.orders %}
        <div class="card mb-4">
          <div class="card-header">
            Order #{{ order.id }}
          </div>
          <div class="card-body">
            <b>Date:</b> {{ order.date }}<br />
            <b>Total:</b> ${{ order.total }}<br />
            <table class="table table-bordered
              table-striped text-center mt-3">
              <thead>
                <tr>
                  <th scope="col">Item ID</th>
                  <th scope="col">Movie</th>
                  <th scope="col">Price</th>
                  <th scope="col">Quantity</th>
                </tr>
              </thead>
              <tbody>
                {% for item in order.item_set.all %}
                <tr>
```

```
            <td>{{ item.movie.id }}</td>
            <td>
              <a class="link-dark"
                href="{% url 'movies.show'
                id=item.movie.id %}">
                {{ item.movie.name }}
              </a>
            </td>
            <td>${{ item.movie.price }}</td>
            <td>{{ item.quantity }}</td>
          </tr>
          {% endfor %}
        </tbody>
      </table>
    </div>
  </div>
  {% endfor %}
    </div>
  </div>
  </div>
</div>
{% endblock content %}
```

Let's explain the previous code:

- We extend the `base.html` template.

- We iterate over each order object stored in `template_data.orders`. For each order, we display its `date` and `total`.

- Then, we iterate we iterate over each item in the current order. The `order.item_set.all` retrieves all related items associated with the current order. For each of those items, we display its `price` and `quantity`, and the corresponding movie id and name.

Adding a link in the base template

Let's add the orders link in the base template. In `/moviesstore/templates/base.html`, in the header section, add the following lines in **bold**:

```
...
{% if user.is_authenticated %}
<a class="nav-link"
  href="{% url 'accounts.orders' %}">Orders
</a>
<a class="nav-link"
```

```
        href="{% url 'accounts.logout' %}">Logout
        ({{ user.username }})
      </a>
      {% else %}
      <a class="nav-link"
        href="{% url 'accounts.login' %}">Login
      </a>
      <a class="nav-link"
        href="{% url 'accounts.signup' %}">Sign Up
      </a>
      {% endif %}
      ...
```

Now, save those files, run the server, and go to `http://localhost:8000/accounts/orders`. If you made a purchase, you would see your corresponding orders (*Figure 12.3*):

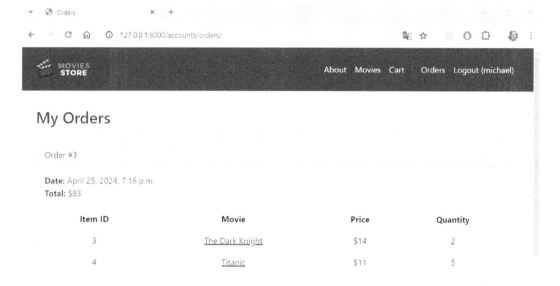

Figure 12.3 – Orders page

We have completed the Movies Store project code. We have implemented all the functionalities planned in *Chapter 1*. Now, let's compare the implemented code with the architecture diagram.

Recapping the Movies Store MVT architecture

The architecture diagram of the Movies Store that we designed in *Chapter 1* served as a blueprint for designing the applications, layers, and code of the Movies Store. We have already implemented all the applications and elements described in that diagram. So, let's quickly recap what we have accomplished so far.

Figure 12.4 displays the complete project tree directory structure and compares it with a simple version of the project architecture. We have successfully implemented four apps (accounts, cart, home, and movies), which contain most of the project's functionalities.

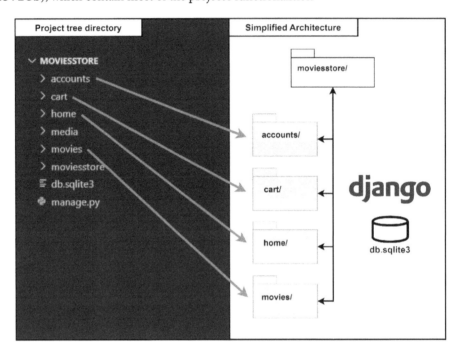

Figure 12.4 – Project tree directory compared with simplified architecture

Figure 12.5 displays the complete architecture. We hope you understand each of the architectural elements better and how they relate to each other.

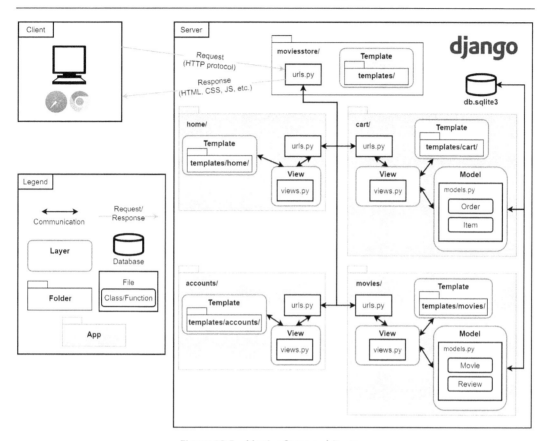

Figure 12.5 – Movies Store architecture

Let's make a last quick analysis:

- We implemented a project-level folder named moviesstore. This folder contained the project-level URL file, which connected with app-level URL files.

- We implemented four Django apps. For each of those apps, we illustrated the communication between the three main layers: models, views, and templates.

- We learned how to divide the code across multiple apps to improve maintainability and separate responsibilities.

- We practiced the implementation of those files and layers by implementing a set of functionalities for our Movies Store project.

What a journey! We've utilized numerous Django modules, libraries, functions, concepts, and elements to implement this project.

Summary

In this chapter, we completed the Movies Store project. We implemented purchase functionality, which took advantage of the `Order` and `Item` models. We created an orders page to allow users to view their orders. We recapped our Movies Store architecture diagram and engaged in comparisons and discussions with the actual project code. We have learned a lot since we started. Now, it's time for the final chapter. Let's learn how to deploy our Movies Store project to the cloud.

13

Deploying the Application to the Cloud

Our project is currently running on our local machine. To make this project accessible to others, we need to deploy it on a server on the internet. A popular way to do this is by deploying our Django project on PythonAnywhere, as it is free to use for small websites. Let's see how to deploy our application to the cloud.

In this chapter, we will be covering the following topics:

- Managing GitHub and Git

- Cloning your code onto PythonAnywhere

- Configuring virtual environments

- Setting up your web app

- Configuring static files

By the end of the chapter, you will have the knowledge and ability to deploy small Python applications on the cloud.

Technical requirements

In this chapter, we will be using Python 3.10+. We will be using Git to upload our code to the cloud, which you can download from `https://git-scm.com/downloads`. Finally, we will be using the **VS Code** editor in this book, which you can download from `https://code.visualstudio.com/`.

The CiA video for this chapter can be found at `https://packt.link/QXahe`

Managing GitHub and Git

To get our code onto sites such as PythonAnywhere, first, we need our code to be on a code-sharing platform such as GitHub or GitLab. In this chapter, we will use GitHub. If you are already familiar with uploading your code to GitHub, please skip the following section and proceed to upload the Movies Store code to a new GitHub repository. Otherwise, you can follow along.

To upload our code to GitHub, we will follow the next steps:

1. Understanding Git and GitHub.
2. Creating a GitHub repository.
3. Uploading our code to GitHub.

Understanding Git and GitHub

Git is a distributed version control system designed to handle everything from small to very large projects with speed and efficiency. It allows multiple developers to collaborate on projects by tracking changes to files (`https://git-scm.com/`).

GitHub is a web-based platform built on top of the Git version control system. It provides hosting for software development projects that use Git for version control (`https://github.com/`).

We'll enhance our Movies Store project to function as a version control system by utilizing Git. Then, we'll host the Movies Store project code on the Cloud using GitHub.

Creating a GitHub repository

A GitHub repository is a central location where files and folders associated with a project are stored and managed. It serves as a version-controlled hub for a project, allowing multiple collaborators to contribute to the development process.

Let's follow the next steps to create a GitHub repository:

1. Go to `https://github.com/` and sign up for an account if you don't have one. Then, create a new repository by clicking on + at the top-right, and select **New repository** (*Figure 13.1*):

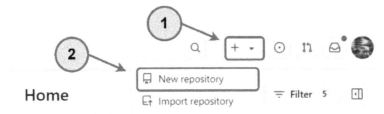

Figure 13.1 – GitHub – create a new repository option

2. Give your repository a name such as moviesstore. Select the **Public** radio box and hit
    Create repository (*Figure 13.2*):

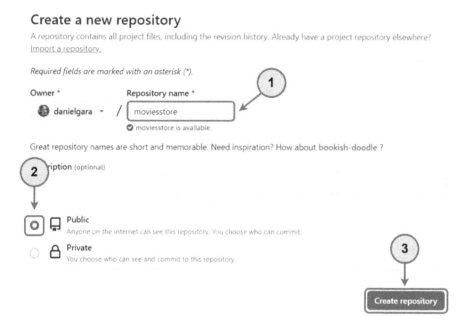

Figure 13.2 – GitHub – creating a new repository

We have successfully created a GitHub repository. We will use it to store the Movies Store project code.
Keep your GitHub repository open; we will use that page in the next section.

Uploading our code to GitHub

We will begin to move our code onto GitHub. In your local machine's Terminal, ensure you have Git
installed by running the following:

```
git
```

If you run the git command in the terminal and see Git usage and commands listed, it indicates that
you have Git installed (*Figure 13.3*):

```
PS C:\Users\yo> git
usage: git [--version] [--help] [-C <path>] [-c <name>=<value>]
           [--exec-path[=<path>]] [--html-path] [--man-path] [--info-path]
           [-p | --paginate | -P | --no-pager] [--no-replace-objects] [--bare]
           [--git-dir=<path>] [--work-tree=<path>] [--namespace=<name>]
           <command> [<args>]

These are common Git commands used in various situations:
```

Figure 13.3 – Executing the git command in the terminal

If you don't see them, you will need to install Git. Visit the Git site (`https://git-scm.com/downloads`) and follow the instructions to install Git. When Git is installed, you might need to close and reopen the Terminal and type "git" in it to ensure that it is installed.

Now that we've installed Git, let's proceed with the next steps to upload our Movies Store project code to our GitHub repository:

1. Open your terminal in the top `moviesstore` folder (the one that contains the `manage.py` file). Then, run the following command:

   ```
   git init
   ```

   The previous command marks your folder as a Git project, allowing you to start tracking changes. A hidden folder named .git is added to the project directory. This folder stores all the metadata, configuration files, and elements that Git needs to track changes and manage the project.

2. Next, run the following command:

   ```
   git add .
   ```

   The previous command adds everything (folders, subfolders, and files) in our project to the staging area, preparing them to be included in the next commit.

3. Then, go ahead and commit the previous changes:

   ```
   git commit -m "first version"
   ```

   The previous command is used to record the changes and inclusions we made to the staging area. When you run git commit, you're essentially creating a snapshot of the current state of your project. You can identify different commits by the descriptive messages you provide.

4. Next, run the following command:

   ```
   git branch -M main
   ```

   This creates a branch called main. This will be the place in which we store our application code.

5. Now, we want to save our Git project on GitHub. In the repository page in GitHub, copy the `git remote add origin <your-origin-path>` command (*Figure 13.4*) and run it in the Terminal (remember to replace `<your-origin-path>` with yours):

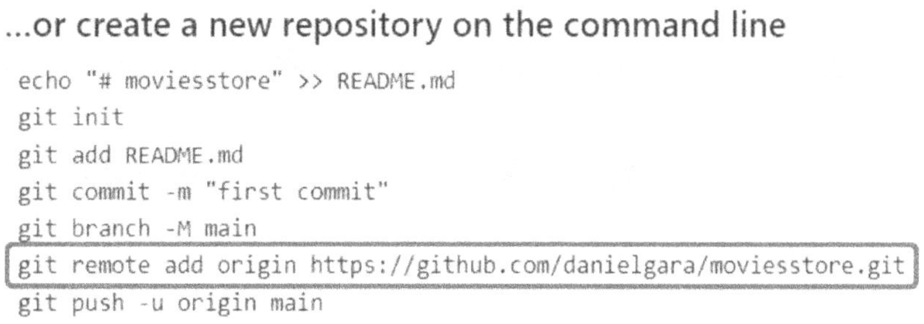

...or create a new repository on the command line

```
echo "# moviesstore" >> README.md
git init
git add README.md
git commit -m "first commit"
git branch -M main
git remote add origin https://github.com/danielgara/moviesstore.git
git push -u origin main
```

Figure 13.4 – Locating your GitHub repository path

`git remote add origin <your-origin-path>`

The previous command is essentially telling Git to create a new remote repository with the name origin and associate it with the URL or path you provide. This will allow you to push your local changes to the remote repository later.

6. To move the code from your local computer to GitHub, run the following:

`git push -u origin main`

If the upload is successful, you should see a message like this (Figure 13.5):

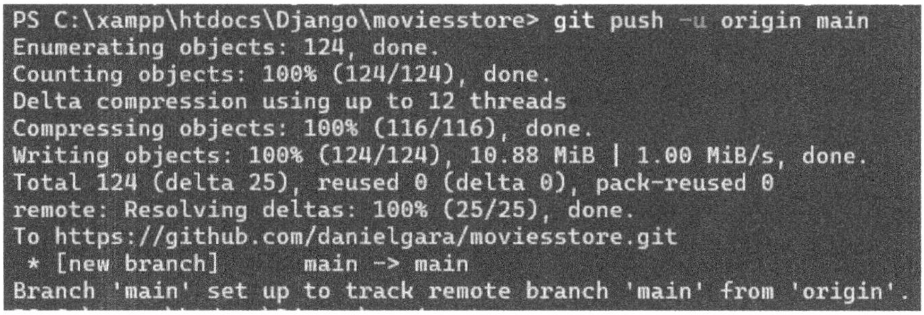

```
PS C:\xampp\htdocs\Django\moviesstore> git push -u origin main
Enumerating objects: 124, done.
Counting objects: 100% (124/124), done.
Delta compression using up to 12 threads
Compressing objects: 100% (116/116), done.
Writing objects: 100% (124/124), 10.88 MiB | 1.00 MiB/s, done.
Total 124 (delta 25), reused 0 (delta 0), pack-reused 0
remote: Resolving deltas: 100% (25/25), done.
To https://github.com/danielgara/moviesstore.git
 * [new branch]      main -> main
Branch 'main' set up to track remote branch 'main' from 'origin'.
```

Figure 13.5 – A successful git push to the GitHub repository

Note

If this is your first time uploading code to GitHub, you will probably see a prompt asking you to log in to GitHub. Please complete that process.

Now, when you reload the GitHub repository page, you should see the Movies Store project structure and files properly uploaded (as shown in *Figure 13.6*):

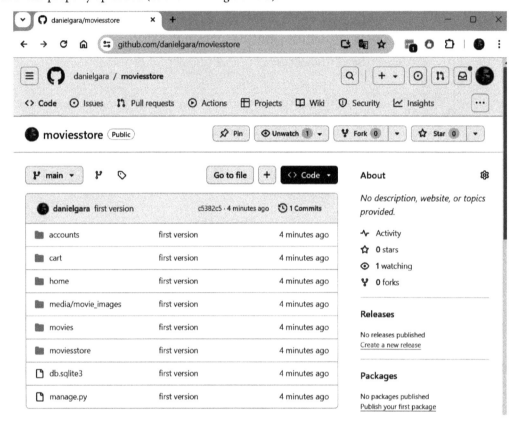

Figure 13.6 – GitHub repository containing the Movies Store project code

> **Note**
>
> Do note that there is much more to Git and GitHub. We have just covered the necessary steps to upload our code to GitHub.

With this, we have now placed our code on GitHub. Next, we will clone it on PythonAnywhere.

Cloning your code onto PythonAnywhere

PythonAnywhere (https://www.pythonanywhere.com/) is a cloud-based platform that provides a web hosting environment for Python applications. It allows users to write, edit, and run Python code directly in their web browser without needing to install any software locally.

The steps to deploy an existing Django project on PythonAnywhere can be found at `https://help.pythonanywhere.com/pages/DeployExistingDjangoProject`, but we'll guide you through them here.

Now that our code is on GitHub, let's proceed with the next steps to create a PythonAnywhere account and move our code from GitHub to PythonAnywhere:

1. Go to `https://www.pythonanywhere.com/registration/register/beginner/` and sign up for a beginner free account if you don't have one.

2. Then, click on **Dashboard | New console | $ Bash** (*Figure 13.7*):

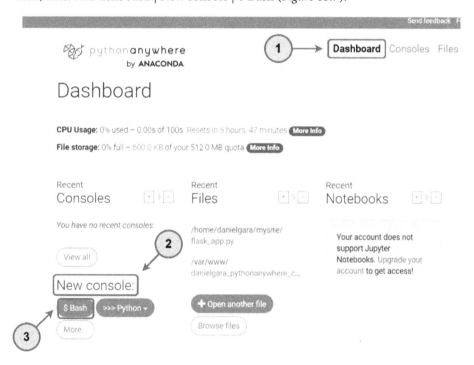

Figure 13.7 – Creating a new console

3.   The previous step will open a Bash console. Back in your GitHub repository, click on **Code** and copy the URL to clone (*Figure 13.8*):

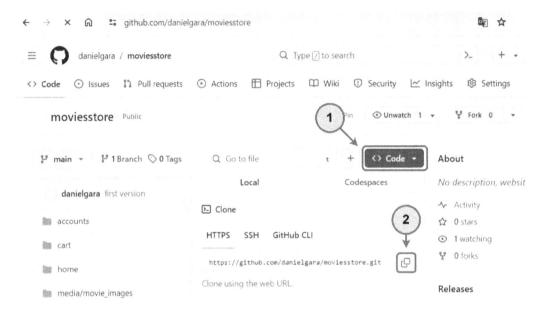

Figure 13.8 – Copying repository URL

4.   To clone the previous repository, go back to the PythonAnywhere Bash console and run the following command (replace the `<repo-url>` part with yours, for example, `git clone https://github.com/danielgara/moviesstore.git`):

```
git clone <repo-url>
```

The previous command takes all of your code from the GitHub repository and clones it in PythonAnywhere. Once the cloning process is complete, you can execute the `ls` command in Bash, and you will see a folder with the repository name containing the repository code (refer to Figure 13.9).

```
15:07 ~ $ git clone https://github.com/danielgara/moviesstore.git
Cloning into 'moviesstore'...
remote: Enumerating objects: 124, done.
remote: Counting objects: 100% (124/124), done.
remote: Compressing objects: 100% (91/91), done.
remote: Total 124 (delta 25), reused 124 (delta 25), pack-reused 0
Receiving objects: 100% (124/124), 10.88 MiB | 30.53 MiB/s, done.
Resolving deltas: 100% (25/25), done.
Updating files: 100% (99/99), done.
15:07 ~ $ ls
README.txt  moviesstore  mysite
15:07 ~ $ 
```

Figure 13.9 – Checking with the ls command that the repository was successfully cloned

We've successfully cloned our repository code into PythonAnywhere. Now, let's configure a virtual environment to be able to run our project.

Configuring virtual environments

A **virtual environment** in Python is a self-contained directory that contains a specific Python interpreter version, along with a set of libraries and packages. It allows you to create an isolated environment for each of your Python projects, ensuring that dependencies are kept separate and do not interfere with each other.

Next, we will create a virtual environment in our PythonAnywhere Bash console to isolate our project code and dependencies. Let's proceed with the following steps:

1. To create a virtual environment in the PythonAnywhere Bash console we have to execute something like this command: `mkvirtualenv -p python3.10 <environment-name>`. For now, we will replace `<environment-name>` with `moviesstoreenv` and run the following:

   `mkvirtualenv -p python3.10 moviesstoreenv`

   We will see the name of virtualenv in Bash, for example, (moviesstoreenv). This means we are in the virtual environment (Figure 13.10):

```
15:19 ~ $ mkvirtualenv -p python3.10 moviesstoreenv
created virtual environment CPython3.10.5.final.0-64 in 32910ms
  creator CPython3Posix(dest=/home/danielgara/.virtualenvs/moviesstoreenv, clear=False, no_vcs_ignore=False, global=
False)
  seeder FromAppData(download=False, pip=bundle, setuptools=bundle, wheel=bundle, via=copy, app_data_dir=/home/danie
lgara/.local/share/virtualenv)
    added seed packages: pip==22.1.2, setuptools==62.6.0, wheel==0.37.1
  activators BashActivator,CShellActivator,FishActivator,NushellActivator,PowerShellActivator,PythonActivator
virtualenvwrapper.user_scripts creating /home/danielgara/.virtualenvs/moviesstoreenv/bin/predeactivate
virtualenvwrapper.user_scripts creating /home/danielgara/.virtualenvs/moviesstoreenv/bin/postdeactivate
virtualenvwrapper.user_scripts creating /home/danielgara/.virtualenvs/moviesstoreenv/bin/preactivate
virtualenvwrapper.user_scripts creating /home/danielgara/.virtualenvs/moviesstoreenv/bin/postactivate
virtualenvwrapper.user_scripts creating /home/danielgara/.virtualenvs/moviesstoreenv/bin/get_env_details
(moviesstoreenv) 15:19 ~ $
```

Figure 13.10 – Bash located in virtualenv

2. Back in our virtualenv, we need to install `django` and `pillow` (as we did in development). So, run the following:

   `pip install django==5.0 pillow`

   The previous execution may take from a couple of minutes to ten minutes. PythonAnywhere has very fast internet, but the filesystem access can be slow, and Django creates a lot of small files during its installation. Thankfully, you only have to do it once. Once it's completed, you should see a message like the one shown in Figure 13.11:

```
(moviesstoreenv) 15:30 ~ $ pip install django==5.0 pillow
Looking in links: /usr/share/pip-wheels
Collecting django==5.0
  Using cached Django-5.0-py3-none-any.whl (8.1 MB)
Requirement already satisfied: pillow in ./.virtualenvs/moviesstoreenv/lib/python3.10/site-packages (10.3.0)
Requirement already satisfied: sqlparse>=0.3.1 in ./.virtualenvs/moviesstoreenv/lib/python3.10/site-packages (from d
jango==5.0) (0.5.0)
Requirement already satisfied: asgiref>=3.7.0 in ./.virtualenvs/moviesstoreenv/lib/python3.10/site-packages (from dj
ango==5.0) (3.8.1)
Requirement already satisfied: typing-extensions>=4 in ./.virtualenvs/moviesstoreenv/lib/python3.10/site-packages (f
rom asgiref>=3.7.0->django==5.0) (4.11.0)
Installing collected packages: django
Successfully installed django-5.0
(moviesstoreenv) 15:39 ~ $
```

Figure 13.11 – Django and pillow installed

We have already configured our virtual environment. For now, you can leave that Bash console open or close it. Now, let's create a web app that utilizes this virtual environment.

Setting up your web app

At this point, we need to be armed with three pieces of information:

- The path to your Django project's top folder (the folder that contains the `manage.py` file). For this project, it is commonly a combination of `/home` and `/<pythonanywhere-user>` and `/<github-repo-name>`. In our case, it was `/home/danielgara/moviesstore`.

- The name of your main project folder (that's the name of the folder that contains your `settings.py` file). In our case, it is `moviesstore`.

- The name of your virtualenv. In our case it is `moviesstoreenv`.

Now, follow the next steps to setting up your web app:

1. In your browser, open a new tab and go to the PythonAnywhere dashboard. Then, click on the **Web** tab and click **Add a new web app** (*Figure 13.12*):

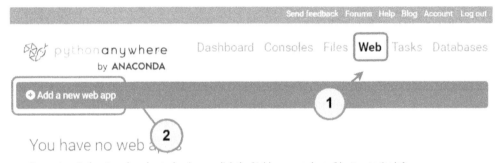

Figure 13.12 – PythonAnywhere Web tab

2. PythonAnywhere will ask you for **Your web app's domain name**. Just click **Next** (*Figure 13.13*):

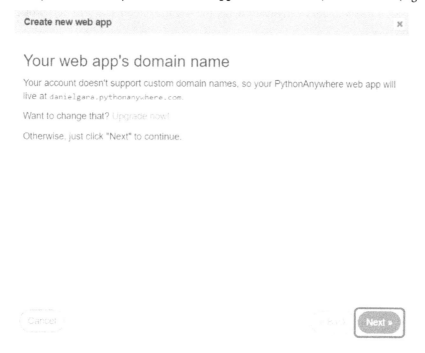

Figure 13.13 – PythonAnywhere domain name

3. In the **Select a Python Web framework** section, choose **Manual configuration** (*Figure 13.14*):

Figure 13.14 – Selecting Manual configuration

> **Note**
>
> Make sure you choose **Manual configuration**, not the **Django** option; that's for new projects only.

4. Select the proper version of Python (the same one you used to create your virtual environment). In our case, it was `Python 3.10` (*Figure 13.15*). Finally, when asked for **Manual configuration**, click **Next** (*Figure 13.16*).

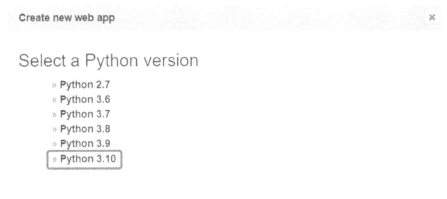

Figure 13.15 – Selecting the right Python version

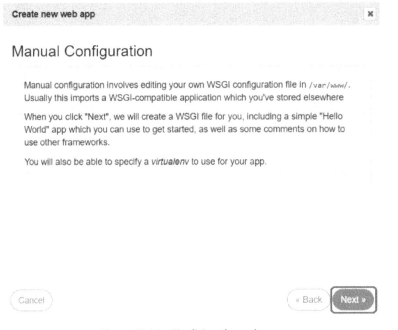

Figure 13.16 – Finalizing the web app

5. Once the web app is created, you need to enter the name of your virtualenv in the **Virtualenv** section (*Figure 13.17*). You can just use its short name, `moviesstoreenv`, and it will automatically complete its full path in `/home/username/.virtualenvs/moviesstoreenv`:

Virtualenv:

Use a virtualenv to get different versions of flask, django etc from our default system ones. More info here. You need to **Reload your web app** to activate it; NB - will do nothing if the virtualenv does not exist.

/home/danielgara/.virtualenvs/moviesstoreenv

Figure 13.17 – Entering the virtualenv name

6. Next, enter the path to your username folder (`/home/<your-username>/`) in the **Code section**, both for **Source code** and **Working directory** (*Figure 13.18*):

Code:

What your site is running.

Source code: /home/danielgara/   Go to directory
Working directory: /home/danielgara/   Go to directory

Figure 13.18 – Entering the path to your code

7. Click the `wsgi.py` file inside the **Code section**, not the one in your local Django project folder (*Figure 13.19*):

Code:

What your site is running.

Source code: /home/danielgara/   Go to directory
Working directory: /home/danielgara/   Go to directory
WSGI configuration file: /var/www/danielgara_pythonanywhere_com_wsgi.py
Python version: 3.10

Figure 13.19 – Accessing the wsgi.py file

This will take you to an editor where you can make changes.

8. Delete everything except the Django section and uncomment that section. Your WSGI file will look something like the following:

```
# +++++++++++ DJANGO +++++++++++
# To use your own django app use code like this:
import os
import sys
```

```
path = '/home/danielgara/moviesstore'
if path not in sys.path:
    sys.path.append(path)

os.environ['DJANGO_SETTINGS_MODULE'] =
    'moviesstore.settings'

from django.core.wsgi import get_wsgi_application
application = get_wsgi_application()
```

Be sure to substitute the correct path to your project, the folder that contains the manage.py file:

```
path = '/home/danielgara/moviesstore'
```

Make sure you put the correct value for DJANGO_SETTINGS_MODULE (where the settings.py file is located):

```
os.environ['DJANGO_SETTINGS_MODULE'] =
    'moviesstore.settings'
```

Finally, save the file.

9. Next, we need to add to the allowed hosts in settings.py. Go to the PythonAnywhere **Files** tab and navigate through the source code directory until you find the settings.py file (*Figure 13.20*):

Figure 13.20 – Accessing the settings.py file

10. Click the `settings.py` file. In `settings.py`, modify the `ALLOWED_HOSTS` variable:

    ```
    ...
    # SECURITY WARNING: don't run with debug turned on in
      production!
    DEBUG = True

    ALLOWED_HOSTS = ['*']
    ...
    ```

    Save the file.

11. Then, go to the **Web** tab and hit the **Reload** button for your domain (*Figure 13.21*):

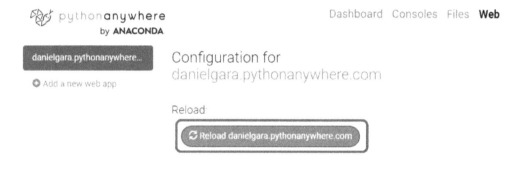

Figure 13.21 – Reloading the web app

The `ALLOWED_HOSTS` settings represent which host/domain names our Django site can serve. This is a security measure to prevent HTTP Host header attacks. We used the asterisk (`*`) wildcard to indicate that all domains are acceptable. In your production projects, you can explicitly list which domains are allowed.

12. Go to your project's URL (it is the blue link in the previous screenshot, for example, `danielgara.pythonanywhere.com`), and the home page should now appear (*Figure 13.22*):

> **Note**
>
> The home page will look strange because we need to configure our application to serve static files (such as images and styles). We will fix it later.

Welcome to the best movie store!!

Figure 13.22 – The PythonAnywhere web app link

We are almost there! Let's fix the static images in the next section.

Configuring static files

Let's fix the problem of our static and media images not appearing:

1. In PythonAnywhere, go back to the **Files** tab and navigate to the settings.py file. We need to add the following in **bold**:

    ```
    ...
    STATIC_URL = 'static/'
    STATIC_ROOT = os.path.join(BASE_DIR, 'static')

    # Default primary key field type
    # https://docs.djangoproject.com/en/4.0/ref/settings/
    #default-auto-field

    DEFAULT_AUTO_FIELD = 'django.db.models.BigAutoField'

    MEDIA_ROOT = os.path.join(BASE_DIR,'media')
    MEDIA_URL = '/media/'
    ...
    ```

The STATIC_ROOT variable defines a central location into which we collect all static files.

2. In PythonAnywhere, go to the **Consoles** tab, and click your **Bash console**. Then, connect to your virtual environment by executing the following command:

    ```
    workon moviesstoreenv
    ```

    Then, go to the moviesstore folder (where the manage.py file is located) by running the following command:

    ```
    cd moviesstore/
    ```

    Execute the following command (Figure 13.23):

    ```
    python manage.py collectstatic
    ```

    This command collects all your static files from each of your app folders (including the static files for the admin app) and from any other folders you specify in settings.py and copies them into STATIC_ROOT:

    ```
    19:29 ~ $ workon moviesstoreenv
    (moviesstoreenv) 19:29 ~ $ cd moviesstore/
    (moviesstoreenv) 19:32 ~/moviesstore (main)$ ls
    accounts  cart  db.sqlite3  home  manage.py  media  movies  moviesstore
    (moviesstoreenv) 19:33 ~/moviesstore (main)$ python manage.py collectstatic

    130 static files copied to '/home/danielgara/moviesstore/static'.
    ```

    Figure 13.23 – Executing the python manage.py collectstatic command

    You need to rerun this command whenever you want to publish new versions of your static files.

3. Next, set up a static file mapping to get our web servers to serve out your static files for you. In the **Web** tab on the PythonAnywhere dashboard, under **Static files**, enter a new record. In **URL**, enter /static/. In **Directory**, enter your project path plus static/, for example, /home/danielgara/moviesstore/static/ (*Figure 13.24*):

Static files:

Files that aren't dynamically generated by your code, like CSS, JavaScript or uploaded files, can be served much faster straight off the disk if you specify them here. You need to **Reload your web app** to activate any changes you make to the mappings below.

URL	Directory	Delete
/static/	/home/danielgara/moviesstore/static/	🗑

Figure 13.24 – Defining the static files

4. Then, in the **Web** tab, hit **Reload**, open your website, and your static images should appear now (*Figure 13.25*):

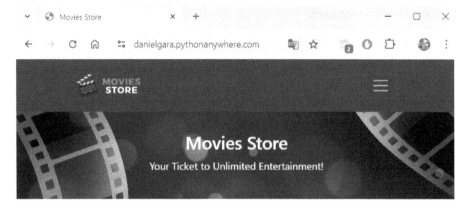

Figure 13.25 – The Movies Store – home page

We did it! Our Movies Store project has been deployed to the cloud. Now you can navigate between the different sections of the website or share your website link with colleagues and friends.

Summary

We have gone through quite a lot of content to equip you with the skills required to create a full-stack Django app. We have covered the major features of Django: templates, views, URLs, user authentication, authorization, models, sessions, forms, and deployment. You now have the knowledge to go and build your own websites with Django. The CRUD functionality in our Reviews app is common in many web applications – for example, you already have all the tools to create a blog, to-do list, or shopping cart web applications.

Hopefully, you have enjoyed this book and would like to learn more from us.

Index

www.packtpub.com

Subscribe to our online digital library for full access to over 7,000 books and videos, as well as industry leading tools to help you plan your personal development and advance your career. For more information, please visit our website.

Why subscribe?

- Spend less time learning and more time coding with practical eBooks and Videos from over 4,000 industry professionals

- Improve your learning with Skill Plans built especially for you

- Get a free eBook or video every month

- Fully searchable for easy access to vital information

- Copy and paste, print, and bookmark content

Did you know that Packt offers eBook versions of every book published, with PDF and ePub files available? You can upgrade to the eBook version at packtpub.com and as a print book customer, you are entitled to a discount on the eBook copy. Get in touch with us at customercare@packtpub.com for more details.

At www.packtpub.com, you can also read a collection of free technical articles, sign up for a range of free newsletters, and receive exclusive discounts and offers on Packt books and eBooks.

Other Books You May Enjoy

If you enjoyed this book, you may be interested in these other books by Packt:

Hands-On Microservices with Django

Tieme Woldman

ISBN: 978-1-83546-852-4

- Understand the architecture of microservices and how Django implements it
- Build microservices that leverage community-standard components such as Celery, RabbitMQ, and Redis
- Test microservices and deploy them with Docker
- Enhance the security of your microservices for production readiness
- Boost microservice performance through caching
- Implement best practices to design and deploy high-performing microservices

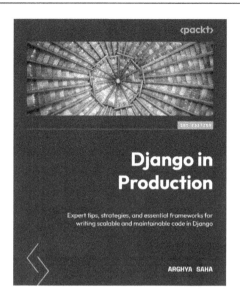

Django in Production

Arghya Saha

ISBN: 978-1-80461-048-0

- Write scalable and maintainable code like a Django expert
- Become proficient in Docker for Django and experience platform-agnostic development
- Explore intelligent practices for continuous integration
- Leverage the power of AWS to seamlessly deploy your application in a production environment
- Optimize unstable systems through effective performance monitoring
- Effortlessly handle authentication and authorization issues
- Automate repetitive tasks by creating custom middleware
- Thoroughly test your code using factory_boy and craft comprehensive API tests

Packt is searching for authors like you

If you're interested in becoming an author for Packt, please visit `authors.packtpub.com` and apply today. We have worked with thousands of developers and tech professionals, just like you, to help them share their insight with the global tech community. You can make a general application, apply for a specific hot topic that we are recruiting an author for, or submit your own idea.

Share Your Thoughts

Now you've finished *Django 5 for the Impatient*, we'd love to hear your thoughts! Scan the QR code below to go straight to the Amazon review page for this book and share your feedback or leave a review on the site that you purchased it from.

`https://packt.link/r/1835461557`

Your review is important to us and the tech community and will help us make sure we're delivering excellent quality content.

Download a free PDF copy of this book

Thanks for purchasing this book!

Do you like to read on the go but are unable to carry your print books everywhere?

Is your eBook purchase not compatible with the device of your choice?

Don't worry, now with every Packt book you get a DRM-free PDF version of that book at no cost.

Read anywhere, any place, on any device. Search, copy, and paste code from your favorite technical books directly into your application.

The perks don't stop there, you can get exclusive access to discounts, newsletters, and great free content in your inbox daily

Follow these simple steps to get the benefits:

1. Scan the QR code or visit the link below

https://packt.link/free-ebook/9781835461556

2. Submit your proof of purchase

3. That's it! We'll send your free PDF and other benefits to your email directly

www.ingramcontent.com/pod-product-compliance
Lightning Source LLC
Chambersburg PA
CBHW080523060326

40690CB00022B/5013